SPEECH AND AUDIO COI WIRELESS AND NETV APPLICATIONS

edited by

Bishnu S. Atal
AT&T Bell Laboratories

Vladimir Cuperman
Simon Fraser University

Allen Gersho
University of California, Santa Barbara

KLUWER ACADEMIC PUBLISHERS
Boston / Dordrecht / London

Distributors for North America:
Kluwer Academic Publishers
101 Philip Drive
Assinippi Park
Norwell, Massachusetts 02061 USA

Distributors for all other countries:
Kluwer Academic Publishers Group
Distribution Centre
Post Office Box 322
3300 AH Dordrecht, THE NETHERLANDS

Library of Congress Cataloging-in-Publication Data

Speech and audio coding for wireless and network applications / edited by Bishnu S. Atal, Vladimir Cuperman, Allen Gersho.
p. cm. -- (The Kluwer international series in engineering and computer science. Communications and information theory)
Includes bibliographical references and index.
ISBN 0-7923-9345-7 (alk. paper)
1. Speech processing systems. 2. Coding theory. 3. Signal processing--Digital techniques. 4. Wireless telecommunication systems. I. Atal, Bishnu S. II. Cuperman, Vladimir. III. Gersho, Allen. IV. Series.
TK7882.S65S6318 1993
621.382'8--dc20 93-13233
CIP

Printed on acid-free paper.

Printed in the United States of America

Speech and audio coding for wireless and network

SPEECH AND AUDIO CODING FOR WIRELESS AND NETWORK APPLICATIONS

THE KLUWER INTERNATIONAL SERIES IN ENGINEERING AND COMPUTER SCIENCE

COMMUNICATIONS AND INFORMATION THEORY

Consulting Editor:
Robert Gallager

Other books in the series:

Digital Communication, Edward A. Lee, David G. Messerschmitt
ISBN: 0-89838-274-2
An Introduction to Cryptology, Henk C.A. van Tilborg
ISBN: 0-89838-271-8
Finite Fields for Computer Scientists and Engineers, Robert J. McEliece
ISBN: 0-89838-191-6
An Introduction to Error Correcting Codes With Applications, Scott A. Vanstone and Paul C. van Oorschot
ISBN: 0-7923-9017-2
Source Coding Theory, Robert M. Gray
ISBN: 0-7923-9048-2
Adaptive Data Compression, Ross N. Williams
ISBN: 0-7923-9085
Switching and Traffic Theory for Integrated Broadband Networks, Joseph Y. Hui
ISBN: 0-7923-9061-X
Advances in Speech Coding, Bishnu Atal, Vladimir Cuperman and Allen Gersho
ISBN: 0-7923-9091-1
Source and Channel Coding: An Algorithmic Approach, John B. Anderson and Seshadri Mohan
ISBN: 0-7923-9210-8
Third Generation Wireless Information Networks, Sanjiv Nanda and David J. Goodman
ISBN: 0-7923-9128-3
Vector Quantization and Signal Compression, Allen Gersho and Robert M. Gray
ISBN: 0-7923-9181-0
Image and Text Compression, James A. Storer
ISBN: 0-7923-9243-4
Digital Satellite Communications Systems and Technologies: Military and Civil Applications, A. Nejat Ince
ISBN: 0-7923-9254-X
Sequence Detection for High-Density Storage Channel, Jaekyun Moon and L. Richard Carley
ISBN: 0-7923-9264-7
Wireless Personal Communications, Martin J. Feuerstein and Theodore S. Rappaport
ISBN: 0-7923-9280-9
Applications of Finite Fields, Alfred J. Menezes, Ian F. Blake, XuHong Gao, Ronald C. Mullin, Scott A. Vanstone, Tomik Yaghoobian
ISBN: 0-7923-9282-5
Discrete-Time Models for Communication Systems Including ATM, Herwig Bruneel and Byung G. Kim
ISBN: 0-7923-9292-2
Wireless Communications: Future Directions, Jack M. Holtzman and David J. Goodman
ISBN: 0-7923-9316-3
Satellite Communications: Mobile and Fixed Services, Michael Miller, Branka Vucetic and Les Berry
ISBN: 0-7923-9333-3

CONTENTS

I INTRODUCTION 1

II LOW DELAY SPEECH CODING 3

1. High Quality Low-Delay Speech Coding at 12 kb/s
J. Grass, P. Kabal, M. Foodeei and P. Mermelstein 5

2. Low Delay Speech Coder at 8 kbit/s with Conditional Pitch Prediction
A. Kataoka and T. Moriya 11

3. Low Delay Coding of Speech and Audio Using Nonuniform Band Filter Banks
K. Nayebi and T. P. Barnwell 19

4. 8 kb/s Low-Delay CELP Coding of Speech
J-H. Chen and M. S. Rauchwerk 25

5. Lattice Low Delay Vector Excitation for 8 kb/s Speech Coding
A. Husain and V. Cuperman 33

III SPEECH QUALITY 41

6. Subjective Assessment Methods for the Measurement of Digital Speech Coder Quality
S. Dimolitsas 43

7. Speech Quality Evaluation of the European, North-American and Japanese Speech Coding Standards for Digital Cellular Systems
E. De Martino 55

8. A Comparison of Subjective Methods for Evaluating Speech Quality
I. L. Panzer, A. D. Sharpley and W. D. Voiers 59

IV SPEECH CODING FOR WIRELESS TRANSMISSION 67

9. Delayed Decision Coding of Pitch and Innovation Signals in Code-Excited Linear Prediction Coding of Speech
H-y. Su and P. Mermelstein 69

10. Variable Rate Speech Coding for Cellular Networks
A. Gersho and E. Paksoy 77

11. QCELP: A Variable Rate Speech Coder for CDMA Digital Cellular
W. Gardner, P. Jacobs and C. Lee 85

12. Performance and Optimization of a GSM Half Rate Candidate
F. Dervaux, C. Gruet and M. Delprat 93

13. Joint Design of Multi-Stage VQ Codebooks for LSP Quantization with Applications to 4 kbit/s Speech Coding
W. P. LeBlanc, S. A. Mahmoud and V. Cuperman 101

14. Waveform Interpolation in Speech Coding
W. B. Kleijn and W. Granzow 111

V AUDIO CODING 119

15. A Wideband CELP Coder at 16 kbit/s for Real Time Applications
E. Harborg, A. Fuldseth, F. T. Johansen and J. E. Knudsen 121

16. Multirate STC and Its Application to Multi-Speaker Conferencing
T. G. Champion, R. J. McAulay and T. F. Quatieri 127

17. Low Delay Coding of Wideband Speech at 32 Kbps Using Tree Structures
Y. Shoham 133

18. A Two-Band CELP Audio Coder at 16 Kbit/s and Its Evaluation
R. D. De Iacovo, R. Montagna, D. Sereno and P. Usai **141**

19. 9.6 kbit/s ACELP Coding of Wideband Speech
C. Laflamme, R. Salami and J-P. Adoul **147**

20. High Fidelity Audio Coding with Generalized Product Code VQ
W-Y. Chan and A. Gersho **153**

VI **SPEECH CODING FOR NOISY TRANSMISSION CHANNELS** **161**

21. On Noisy Channel Quantizer Design for Unequal Error Protection
J. R. B. de Marca **163**

22. Channel Coding Schemes for the GSM Half-Rate System
H. B. Hansen, K. J. Larsen, H. Nielsen and K. B. Mikkelsen **171**

23. Combined Source-Channel Coding of LSP Parameters Using Multi-Stage Vector Quantization
N. Phamdo, N. Farvardin and T. Moriya **181**

24. Vector Quantization of LPC Parameters in the Presence of Channel Errors
K. K. Paliwal and B. S. Atal **191**

25. Error Control and Index Assignment for Speech Codecs
N. B. Cox **203**

VII **TOPICS IN SPEECH CODING** **209**

26. Efficient Techniques for Determining and Encoding the Long Term Predictor Lags for Analysis-by-Synthesis Coders
I. A. Gerson and M. A. Jasiuk **211**

27. Structured Stochastic Codebook and Codebook Adaptation for CELP
T. Taniguchi, Y. Tanaka and Y. Ohta 217

28. Efficient Multi-Tap Pitch Prediction for Stochastic Coding
D. Veeneman and B. Mazor 225

29. QR Factorization in the CELP Coder
P. Dymarski and N. Moreau 231

30. Efficient Frequency-Domain Representation of LPC Excitation
S. K. Gupta and B. S. Atal 239

31. Product Code Vector Quantization of LPC Parameters
S. Wang, E. Paksoy and A. Gersho 251

32. A Mixed Excitation LPC Vocoder with Frequency-Dependent Voicing Strength
A. V. McCree and T. P. Barnwell III 259

33. Adaptive Predictive Coding with Transform Domain Quantization
U. Bhaskar 265

34. Finite-State VQ Excitations for CELP Coders
A. Benyassine, H. Abut and G. C. Marques 271

AUTHOR INDEX 277

INDEX 279

SPEECH AND AUDIO CODING FOR WIRELESS AND NETWORK APPLICATIONS

PART I

INTRODUCTION

In recent years, new applications in digital wireless and network communication systems have emerged which have spurred significant developments in speech and audio coding. Important advances in algorithmic techniques for speech coding have recently emerged and resulted in systems which provide high quality digital voice at bit rates as low as 4 kbit/s. Significant advances in low-rate speech coding has been achieved as a result of the new requirements defined for half-rate digital cellular communications, personal communications networks, and other low rate applications. Progress in low-delay speech coding recently resulted in the CCITT G.728 16 kbit/s speech coding standard and in the preliminary work for the future CCITT 8 kbit/s standard. Increasing attention is also being given today to audio coding (including, in particular, wideband speech). Advances in programmable signal processor chips have kept pace with the increasing complexity of the more recent coding algorithms. The rapid technology transfer from research to product development continues to keep the pressure on speech coding researchers to find better and more efficient algorithms to meet the demanding objectives of the users and standards organizations. In particular, low-rate voice technology is converging with the needs of the rapidly evolving digital telecommunication networks.

The pace and scope of activity in speech coding was evident to attendees of the second IEEE Workshop on Speech Coding for Telecommunications held in Whistler, British Columbia, Canada, in September 1991. Thus, we felt it would be of value to publish a book that contains a cross-section of the key contributions in speech and audio coding that have emerged in the past two years, providing a useful sequel to the book *Advances in Speech Coding* which we edited two years ago (Kluwer Academic Publishers, 1991). We invited a selection of key contributors to the field, most of whom gave papers at the Whistler workshop, to contribute a chapter to this book based on their recent work in speech or audio coding. The focus was limited to topics of relevance to wired or wireless telecommunication networks. Each submitted contribution was subjected to a peer review process to ensure high quality.

This volume contains 34 chapters, loosely grouped into six topical areas. The chapters in this volume reflect the progress and present the state of the art in low bit rate speech coding primarily at bit rates from 2.4 kbit/s to 16 kbit/s. Together they represent important contributions from leading researchers in the speech coding community.

The book contains papers describing technologies that are under consideration as standards for such applications as digital cellular communications (the half-rate American and European coding standards). The book includes a section on the important topic of speech quality evaluation. A section on audio coding covers not only 7 kHz bandwidth speech but also wideband coding applicable to high fidelity music. One of the sections is dedicated to low-delay speech coding, a research direction which emerged as a result of the CCITT requirement for an universal low-delay 16 kbit/s speech coding technology and now continues with the objective of achieving toll quality with moderate delay at a rate of 8 kbit/s. A significant number of papers address future research directions. We hope that the reader will find the contributions instructive and useful.

We would like to take this opportunity to thank all the authors for their contributions to this volume, for making revisions as needed based on the reviews, and for meeting the very tight deadlines. We wish to thank Kathy Cwikla, at Bell Laboratories, Murray Hill for her valuable help in compiling the material for this volume.

Bishnu S. Atal
Vladimir Cuperman
Allen Gersho

PART II

LOW DELAY SPEECH CODING

Speech coders have traditionally been characterized on the basis of three primary criteria: quality, rate, and implementation complexity. Recently *delay* has also become an important specification for many applications. A very stringent delay objective for network applications, led to the development of the 16 kb/s LD-CELP algorithm, with "toll" quality and a one-way coding delay of only 2 ms. This algorithm has been recently adopted as CCITT Recommendation G.728. Subsequent interest has focused on the increasingly difficult challenge of obtaining the same high quality at lower bit rates. In this section, five papers offer a cross-section of more recent efforts to advance the state-of-the-art in low delay coding. Grass et al. examine and compare CELP and tree structures for low delay 12 kb/s coding. Kataoka and Moriya describe an 8 kb/s low delay CELP coder with a novel long delay predictor configuration. Chen and Rauchwerk present a low delay CELP coder at 8 kb/s which includes interframe coding of the pitch. Another 8 kb/s low delay CELP coder with lattice short delay prediction is described by Husain and Cuperman with a comparison of forward and backward options for long delay prediction. Nayebi and Barnwell consider low delay sub-band coding with nonuniform filter banks with a technique that reduces delay while avoiding any noticeable degradation in the reconstruction.

1

HIGH QUALITY LOW-DELAY SPEECH CODING AT 12 KB/S

J. Grass [1], P. Kabal [1,2], M. Foodeei [2] and P. Mermelstein [1,2,3]

[1] INRS-Télécommunications
Université du Québec
Verdun, Quebec
Canada H3E 1H6

[2] Electrical Engineering
McGill University
Montreal, Quebec
Canada H3A 2A7

[3] BNR
16 Place du Commerce
Verdun, Quebec
Canada H3E 1H6

INTRODUCTION

For low-delay speech coders, the research challenge is to obtain higher compression rates while maintaining very high speech quality and meeting stringent low delay requirements. Such coders have applications in telephone networks, mobile radio, and increasingly for in-building wireless telephony.

A low-delay CELP algorithm operating at 16 kb/s has been proposed for CCITT standardization [1, 2, 3, 4]. An alternate coding structure operating at the same rate is based on an ML-Tree algorithm [5]. Both algorithms offer near-network quality with coding delays below 2 ms at 16 kb/s. In this work, we modify these basic coder structures to operate at the reduced rate of 12 kb/s while retaining high speech quality.

In the low-delay coders considered here, the following common features may be identified.

- excitation selection using analysis-by-synthesis,
- high performance predictors for redundancy removal,
- gain scaling and adaptation,
- perceptual weighting (noise-shaping), and
- innovation sequence or codebook with delayed decisions

Delayed-decision coding, as implemented in codebook (CELP), tree, and trellis coding, can efficiently represent the residual signal. This is done by postponing the decision as to which quantized residual signal is to be selected. In an analysis-by-synthesis approach, the search for the optimum excitation dictionary or codebook entry at the encoder is effectively obtained by systematically examining the performance resulting from the use of each sequence. The sequence with the lowest perceptually weighted error (original signal sequence to reconstructed signal) is selected. To generate the reconstructed signal, the encoder uses a replica of the decoder. The index corresponding to the selected sequence entry is transmitted to the decoder. In addition, adaptive gain scaling of the excitation signal is used since it improves the excitation representation by reducing the dynamic range of the excitation set. At the encoder, the error signal is passed through a perceptual weighting filter prior to the error minimization. At the decoder, an optional postfiltering stage can be added to further improve perceptual quality.

Assuming a sampling rate of 8 kHz, the low-delay requirement for network applications limits the encoder delay to 5–8 samples (0.625–1.0 ms). The back-to-back delay for an encoder/decoder is usually 2–3 times the encoder delay. This meets the objective of 2 ms. The overall coder bit-rate is obtained by multipling the sampling frequency f_s by the number of bits/sample ($I = f_s \times R$).

For block-based coding, if a coder sequence (R bits/sample) of length N and a codebook size of J are used, the following relation holds.

$$R = \frac{1}{N} \log_2 J = \frac{k}{N} \quad (J = 2^k). \tag{1}$$

Fractional coding rates are easily obtained by selecting the proper codebook size J and codevector dimension N.

An alternative to block-based coding is a sliding window code for the excitation. In tree and trellis coding, different sequences have several common elements and individual sequences form a path in the tree or trellis. Tree structures [6, 7] are considered here. A consistent assignment of branch number is used throughout the tree which results in a unique *path map* for each path sequence. The path information for the best path is transmitted to the decoder. The number of branches b, per node is called the *branching factor*. If β symbols per node are used, the encoding rate R in bits per symbol is given by

$$R = \frac{1}{\beta} \log_2 b = \frac{k}{\beta} \quad (b = 2^k). \tag{2}$$

Fractional rates can be achieved either by selecting a β value greater than one (*multi-symbols/node*) or by using the concept of a *multi-tree*. In the latter alternative, the branching factor of the tree at different depths changes along the paths (see [8, 9] for more detail).

LOW-DELAY BLOCK-BASED CODING

The low-delay CELP algorithm originally designed for 16 kb/s [2], was modified to operate at 12 kb/s. The bit-rate of the block-based coder is determined by the sampling rate multiplied by the codebook size (number of bits) and divided by the vector length used in the codebook (Eqn. 1). The sampling rate was kept fixed at 8 kHz. A number of different combinations of the parameters were examined. The best of these combinations was found to be a 9 bit codebook and a 6-sample vector size (which corresponds to an encoding delay of 0.75 ms). The codebook design uses a full search approach rather than partitioning into shape/gain sub-codebooks. The codebook was retrained for the lower bit-rate.

The modified coder operating at 12 kb/s maintains good quality for female talkers but the quality degrades somewhat for male speakers. This difference can be attributed to the ability of the 50th order predictor (autocorrelation with analysis updated every 24 samples) to capture some aspects of pitch for female talkers but not for male talkers. Higher order predictors were studied by Foodeei and Kabal [9, 10]. High order (up to 80) covariance analysis allows for the capture of pitch redundancies associated with male talkers. Furthermore, the Cumani algorithm provides a numerically stable algorithm for determining the coefficients of the high-order filter [11].

Using the covariance-lattice predictor in the block-based coder at 12 kb/s instead of the autocorrelation predictor, the quality of the male speech is improved. The covariance-lattice predictor has been shown to increase prediction gain over 2 dB for male speakers [10]. In the 12 kb/s coder, the overall objective performance of the coder in terms of SNR did not change. This may be attributed to the the fact that the adaptation is based on the reconstructed speech. Perceptually however, the covariance-lattice technique provides improvements in the coder for male speakers.

LOW-DELAY TREE CODER

The ML-Tree algorithm was originally used in a configuration with a 3-tap pitch predictor. The adaptive predictor, with dynamic determination of the pitch lag, suffers from error propagation effects. Using an 8th order formant predictor and a simple gain adjustment procedure, the ML-Tree coder at 16 kb/s has speech comparable to that of LD-CELP at the same bit rate [9, 12].

At 16 kb/s, the coding tree has a branching factor of 4 at each sample (2 bits per sample). Our strategy to lower the bit rate is to use combined vector-tree coding (*multi-symbols/node*). The encoding delay is a function of the path length and the number of samples populating each node. The overall bit-rate is given by the sampling rate divided by the number of samples considered at each node and multiplied by the number of bits to represent the branching factor (Eqn. 2). Two configurations were studied, one using 3 bits for the branching factor and 2 samples per node while in the second configuration 6 bits are used for the branching factor and 4 samples per node. The former combination was preferred.

Prediction Filter

The original implementation of the low-delay tree coder uses the generalized predictive coder configuration [5]. In this structure, the reconstruction error is given by $R(z) = Q(z)\frac{1-N_1(z)}{1-F(z)}$. $F(z)$ is the predictor filter, $N_1(z)$ is the noise feedback function and $Q(z)$ is the quantization error. $N_1(z)$ is set equal to $F(z/\mu_1)$. The feedback filter in the this structure provides a method to shape the noise spectrum.

An alternative configuration of the generalized predictive coder structure is that given by Atal and Schroeder [13]. In this closed-loop structure shown in Fig. 1, the perceptual weighting takes the same form as that used in the block-based coder; $W(z) = \frac{1-N_2(z)}{1-N_1(z)}$ where $N_1(z)$ is set equal to $F'(z/\mu_1)$ and $N_2(z)$ to $F'(z/\mu_2)$. The noise feedback filter is no longer directly linked to the prediction filter. The weighting filter can be determined from the clean input speech signal. Furthermore, the prediction filter and perceptual filter need not be of the same order. The noise feedback filters were 10th order filters, adapted

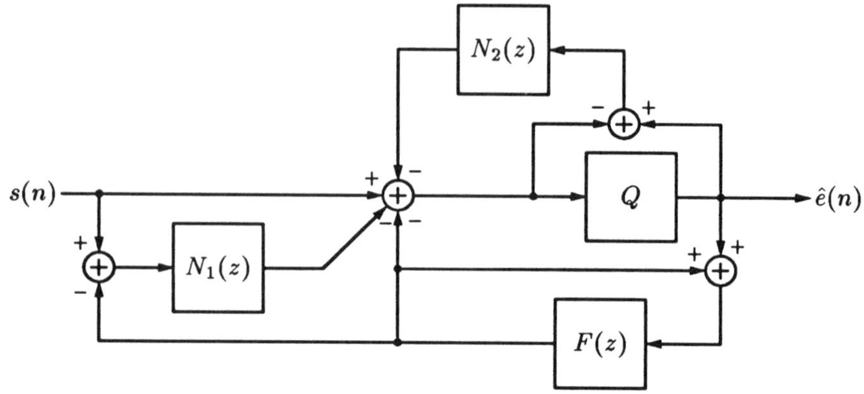

Fig. 1 A closed-loop configuration with generalized noise feedback

from the clean speech (the same as in LD-CELP). For this choice, the resulting speech was significantly better than that for the original configuration of the low-delay tree coder.

For the prediction filter, a configuration using a high order covariance filter and a configuration using a separate pitch filter were compared. The separate pitch filter performed better in terms of reduced pitch spikes in the residual, but subjectively there was little difference.

Gain Adapter

Several gain adaptation schemes were evaluated in the context of the low-delay tree coder. Particular attention was given to the adaptive logarithmic gain update strategy originally used in the 16 kb/s LD-CELP. It was found that the simple gain adaptation scheme proposed by Iyengar [5] achieved SNR results similar to the more complex gain adapters. Perceptually, a slight preference is given to the LD-CELP gain update method.

Dictionary Training

The dictionary for the innovation tree of the coder can be populated in a random fashion [5]. However, improvements as large as 1.5 dB in the performance of the coder at 12 kb/s were achieved by a new training procedure of the dictionary (training speakers and sentences were different than those used for testing).

The training procedure initially uses a randomly populated codebook. In each iteration, the coder is run, accumulating the unquantized prediction errors (residuals) associated with each released node of the tree. Each unquantized residual is assigned to a Voronoi cell corresponding to an entry in the dictionary with smallest distance to this residual. Note that due to the delayed nature of the tree coder, the unquantized residuals must be retained for the length of the delay. Further, the gain value used at each node of the tree must be kept so that the unquantized residual can be appropriately scaled. The centroid of the unquantized residuals in each Voronoi cell is found and used to replace the associated dictionary entry in the previous iteration. With an updated dictionary, this process is repeated for several iterations.

DISCUSSION

The speech quality for the block-based coder operating at 12 kb/s is remarkably good. The principal difference when compared to 16 kb/s LD-CELP is a modest degradation for some male speakers. In comparing the two coders at 12 kb/s, the low-delay tree coder produces speech that is slightly better perceptually than that of the block-based coder.

We noted a significant improvement in the low-delay tree coder with the change to a generalized perceptual weighting, with the weighting filter determined from the clean speech rather than the reconstructed speech. Further work is warranted to compare the noise feedback as used in the tree coder with the open-loop weighting used in block-based coders. In addition, the use of high order covariance-lattice predictors in tree coders needs further investigation.

REFERENCES

1. AT&T contributions to CCITT Study Group XV and T1Y1.2 (October 1988–July 1989).
2. Detailed description of AT&T's LD-CELP algorithm, contributions to CCITT Study Group XV, Nov. 1989.
3. Draft recommendation G.728 (coding of speech at 16 kb/s using LD-CELP), CCITT Study Group XV, Dec. 1991.
4. J.-H. Chen, "High-quality 16 kb/s speech coding with a one-way delay less than 2 ms", *Proc. Int. Conf. on Acoust. Speech, Signal Processing*, (Albuquerque, NM), April 1990, pp. 453–456.
5. V. Iyengar and P. Kabal, "A low-delay 16 kbits/sec speech coder", *IEEE Trans. Signal Processing*, vol. 39, May 1991, pp. 1049–1057.
6. J. B. Anderson and J. B. Bodie, "Tree coding of speech," *IEEE Trans. on Inform. Theory*, vol IT-21, pp. 379–387, July 1975.
7. N. S. Jayant and P. Noll, *Digital Coding of Waveforms*, Prentice-Hall, 1984.
8. J. D. Gibson and W.-W. Chang, "Fractional rate multi-tree speech coding," *IEEE Trans. Commun.*, vol. 39, pp. 963–974, June 1991.
9. M. Foodeei, "Low-delay speech coding at 16 kb/s and below", *Master of Eng. Thesis , Dept. of Elect. Eng. McGill University*, (May 1991).
10. M. Foodeei and P. Kabal, "Backward adaptive prediction: high-order predictors and formant-pitch configuration", *Proc. Int. Conf. on Acoust. Speech, Signal Processing*, (Toronto, Canada), May 1991, pp. 2405–2408.
11. A. Cumani, "On a covariance lattice algorithm for linear prediction", *Proc. Int. Conf. on Acoust. Speech, Signal Processing*, (Paris, France), 1982, pp. 651–654.
12. M. Foodeei and P. Kabal, "Low-delay CELP and Tree coders: comparisons and performance improvements", *Proc. Int. Conf. on Acoust. Speech, Signal Processing*, (Toronto, Canada), May 1991, pp. 25–28.
13. B. S. Atal and M. R. Schroeder, "Predictive Coding of Speech Signals and Subjective Error Criteria", *IEEE Trans. Acoust. Speech, Signal Processing*, vol. ASSP-27, June 1979, pp. 247–254.

2

LOW DELAY SPEECH CODER AT 8 kbit/s WITH CONDITIONAL PITCH PREDICTION

Akitoshi Kataoka and Takehiro Moriya

NTT Human Interface Laboratories
Musashino-Shi, Tokyo, 180 Japan

INTRODUCTION

Medium bit-rate speech coding has been receiving much attention for use in communication systems [1,2,3]. In North America, Japan, and Europe, a lot of research has been carried out for digital cellular radio systems at around 8 kbit/s. Real-time high-quality coders have been built using DSP chips [4,5]. All these systems, however, require echo cancellers because the coding delays are 50 to 100 ms. Without echo control devices, communication might be disrupted by echoes reflected from the hybrid circuit of the receiving telephones.

The delay between the speech coder and decoder should be as small as possible. At present, medium bit-rate, low-delay speech coders are the key to improving overall communication quality [6,7]. However, conventional speech waveform coders are either low-bit-rate long-delay or high-bit-rate low-delay. CELP [8,9] and VSELP [4] based on frame-wise processing fall into the first category, and delta modulation and ADPCM [10] based on sample-by-sample processing fall into the second. There has been no method to bridge these two categories.

This paper reports on a low-delay 8-kbit/s coder. It uses a conditional pitch prediction scheme instead of forward pitch prediction, backward-adaptive gain quantization (gain for pitch component and gain for random component) and a switchover mechanism for the synthesis filter. The design is described and the SNR improvement of each technique is shown. The quality is evaluated by pair-comparison tests with 5/6/7-bit μ-law PCM coders.

CODER OUTLINE

The proposed coder [11], (Fig. 1), is based on the backward-adaptive CELP coder. The coder has two excitation sources for the LPC synthesis filter. One is a pitch component, the other is a random component. The encoder chooses the best excitation source which minimizes the perceptually weighted distortion between the input speech and the synthesized speech.

A pitch candidate is obtained by backward analysis from the residual signal. The codebook is either a random codebook or a trained codebook. The

LPC synthesis filter is estimated only from the reconstructed signal in order to achieve a low delay. Pitch gain quantization is controlled by the correlation coefficient of the residual signal in the previous frame. The proposed coding system uses an excitation gain quantizer with the same adaptation rule as used in LD-CELP. That is, the excitation gain is predicted by the logarithmic gain sequence of previously quantized-and-scaled excitation vectors [12]. The perceptual weighting filter is ARMA type.

The transmitted parameters are pitch delay (preselected by backward analysis), pitch gain (controlled by the autocorrelation coefficient), excitation gain (backward-adapted), the excitation shape code, and side information for filter selection (described in switchover of the synthesis filter).

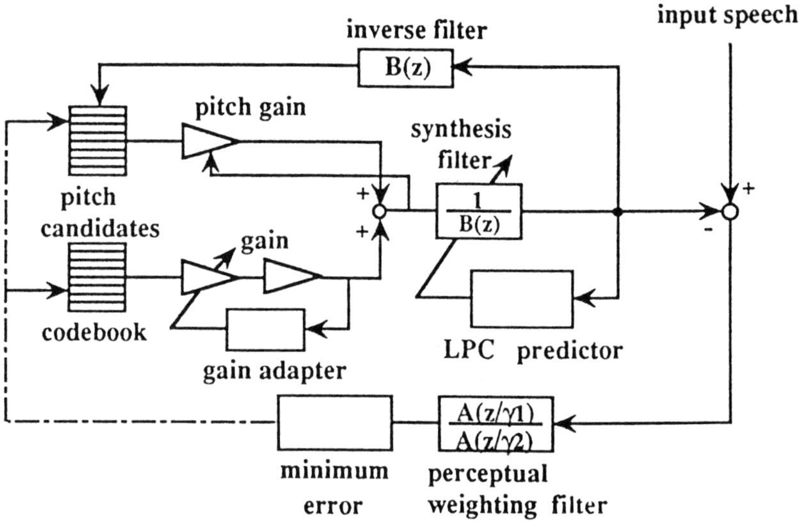

Fig.1 Proposed coder

SHORT-TERM PREDICTION

The proposed coder extracts LPC parameters from the windowed reconstructed signal using backward linear prediction. Two LPC analyses are needed at the encoder, as shown in Fig. 1. One finds the coefficients of the synthesis filter, which are used in both the encoder and the decoder. The p-th order all-pole filter (p=16) is represented by $1/B(z)$, where

$$B(z) = 1 - \sum_{i=1}^{p} b_i z^{-i} \tag{1}$$

The other LPC analysis is for the perceptual weighting filter used only in the encoder [13]. The q-th order all-pole filter is represented by $1/A(z)$, and the weighting filter is $H(z)$ with noise shaping factors γ_1 and γ_2. $A(z)$ and $H(z)$ are given by

$$A(z) = 1 - \sum_{i=1}^{q} a_i z^{-i} \tag{2}$$

$$H(z) = \frac{A(z/\gamma_1)}{A(z/\gamma_2)} \tag{3}$$

$A(z)$ is estimated from the current speech signal, whereas, $B(z)$ is estimated only from the reconstructed signal in order to achieve a low delay. Since the reconstructed signal is available in both the encoder and decoder, the LPC parameters are not transmitted, so there is no limit on the number of parameters used. However, to avoid interfering with the pitch prediction, the order of the LPC is limited to 16. Both LPC analyses are performed by the autocorrelation method with half the Hamming window. After a trial listening test, the perceptual weighting filter was given q=16, γ_1=0.9, and γ_2=0.5.

EXCITATION CODEBOOK

In conventional CELP, the excitation vector is selected from a random codebook. A structured codebook can improve the coder performance in terms of quality, complexity, and robustness against channel errors. With backward-adaptive prediction, a structured or trained codebook is more important because the excitation signal should have some variations in frequency spectrum to compensate for the response of the synthesis filter being different from the ideal LPC synthesis filter. The power spectrum of a synthesis filter derived from quantized speech tends to be contaminated by a quantization noise especially at higher frequencies. Therefore a simple low-pass noise codebook can provide better SNR and quality than a white noise codebook. When speech changes from silence to voice, excitation vector should have some pulsive samples.

The proposed coder uses a trained codebook generated by the generalized Lloyd algorithm [14] within a closed loop of the encoding process. This means that the distortion measure for both finding the code and calculating the centroids is identical to the one used in the encoding process. A trained codebook improves the SNR and quality even more than the low-pass noise codebook.

CONDITIONAL PITCH PREDICTION

Long-term or pitch-period prediction improves speech quality. In forward CELP, parameters such as pitch delay and tap gain are updated every 5-ms subframe. In the proposed coder, frame lengths should be less than 3 ms. In this case, a smaller number of bits represents the pitch information. However, we can make use of the fact that a shorter frame length will have a more highly correlated pitch parameter.

Conditional pitch prediction reduces both the transmission bit-rate and the computational complexity without losing performance. The following three steps are used for finding the pitch lag:

1) Open-loop backward analysis selects M (out of N) lag candidates.

2) A non-integer delay single-tap predictor [15] is applied to only the selected candidates. The new lag candidates are four times as precise as the ones in step 1.

3) Closed-loop forward analysis finds the best lag out of the candidates.

The open-loop analysis calculates the autocorrelation of the residual signal of the reconstructed signal, and selects the M lags with the highest correlation. The reconstructed signal is available in both the encoder and decoder. This means that only $log_2(M)$ bits, instead of $log_2(N)$ bits, have to be sent from the encoder to the decoder. This coder uses N=128 and M=16, so three bits are saved in representing the pitch delay. A non-integer delay is applied to the selected candidates. Closed-loop analysis finds the best pitch lag candidate by comparing the synthesized signal and the speech. Backward analysis uses only the reconstructed signal, while forward analysis uses the current speech frame.

BACKWARD ADAPTIVE GAIN

The proposed coder uses a backward pitch gain adaptation. When the correlation coefficient of the residual signal is large, the optimum pitch gain is concentrated to unity. When the correlation is small, the optimum pitch gain extends from 0 to 2.0. Based on this observation, the step size of the pitch gain quantization is adapted by the correlation coefficient of the residual signal of the previous frame.

SWITCHOVER OF THE SYNTHESIS FILTER

Since pitch-adaptive excitation and random excitation are determined sequentially, a new set of prediction coefficients is available for random excitation. The synthesis filter $1/B_0(z)$ is determined using a previously reconstructed signal. Optimum pitch parameters are determined according to $1/B_0(z)$. Then the partially synthesized signal is generated at the current frame in order to obtain a new set of LPC parameters for $1/B_1(z)$. The new autocorrelation is calculated by using extended window which contains both the previously reconstructed signal and the partially reconstructed signal in the current frame. A new set of LPC parameters contains the newest information of the current frame. The switchover of the synthesis filter is shown in Fig. 2. When there is no good pitch candidate, new LPC parameters are not always desirable. This is a problem of whether an old or a new synthesis filter is better. There are two methods of selecting the synthesis filter.

A. Select the synthesis filter based on the information available at the decoder.

B. Select the synthesis filter that provides the minimum distortion between the input speech and the synthesized speech.

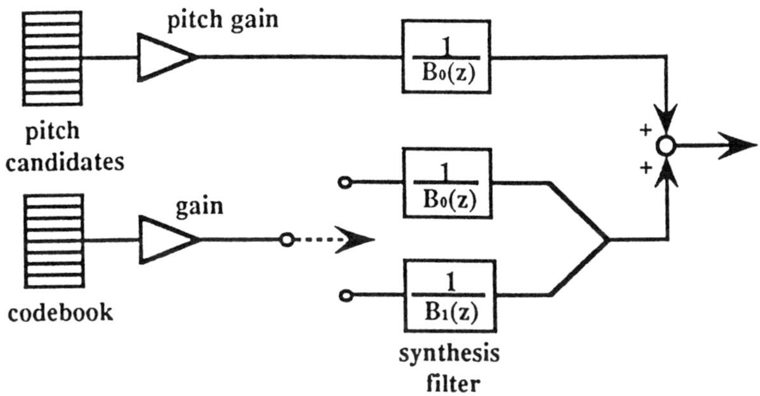

Fig.2 Switchover of the synthesis filter

Type **A** does not need side information. We found the normalized residual powers of $1/B_0(z)$ and $1/B_1(z)$, d_0 and d_1, to be reasonable measures for the selection, where

$$d = \prod_{i=1}^{p} (1 - k_i^2) \tag{4}$$

k_i: PARCOR coefficient

We use the filter whose d is smaller, since smaller quantization distortion is expected if d is small. Type **B** needs one bit of side information for selecting the filter, so it needs additional computation of distortion.

PERFORMANCE EVALUATION

Performance improvements due to the conditional pitch prediction, non-integer delay, adaptive pitch gain quantization, switchover of the synthesis filter, and the trained codebook were evaluated. Table 1 shows the bit allocation of the proposed coders. The results are shown in Fig. 3. The SNR values were averaged over 14 short Japanese sentences (spoken by 5 female, 5 male and 4 children), none of which were in the training sequence of the excitation codebook. Note that the bit-rate is fixed at 8 kbit/s by setting the vector dimension equal to T, the number of bits per frame or vector. The pitch period was set to be longer than the frame length. Each coder is summarized below.

A: A conventional backward-CELP with forward pitch prediction.

B: Conditional pitch prediction

C: Pitch delay is four times as precise as that of **B**.

D: Backward pitch gain adaptation

E: Switchover of the synthesis filter (type **A**)

F: Switchover of the synthesis filter (type **B**)

G: Identical codec to **F** with a trained codebook
The others (from **A** to **F**) use a random codebook.

	A	B	C	D	E	F	G
Pitch lag (bits)	7	4	4	4	4	4	4
Pitch gain (bits)	2	2	2	2	2	2	2
Non-integer (bits)	–	–	2	2	2	2	2
Codebook shape (bits)	10	10	10	10	10	10	10
Codebook gain (bits)	4	4	4	4	4	4	4
Filter selection (bit)	–	–	–	–	–	1	1
Total (bits)	23	20	22	22	22	23	23
Frame length (samples)	23	20	22	22	22	23	23

Table 1 Coding bits for each coder

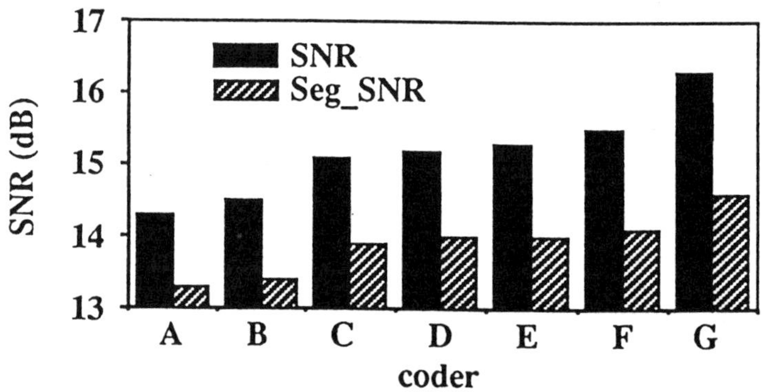

Fig.3 Performance of each each coder

The conditional pitch prediction improved the SNR by 0.2 dB. Non-integer delay was only applied to the final candidate pruned by the conditional pitch prediction. Non-integer delay improved the SNR by 0.6 dB. These schemes are especially useful for female and children's speech. Backward-adaptive quantization of pitch gain also improves the SNR by a simple operation. The switchover of the synthesis filter is useful for backward-adaptive prediction. Although the computational complexity increases, type **B** (switched by side information) is better than type **A** (switched by normalized prediction error). These techniques improved the SNR by 0.4 dB. Finally, the trained codebook further improved the SNR by 0.8 dB. Overall, the proposed coder achieved a performance of 16.3 dB. This SNR is 2 dB better than with the conventional backward-CELP coding.

The quality of the speech coded by method **G** was compared with 5/6/7-bit μ-law PCM in pair-comparison tests. Listeners were six trained females. The results are shown in Fig. 4. In all cases, the quality was superior to that of 6-bit μ-law PCM. The quality of female speech is equivalent to 7-bit PCM.

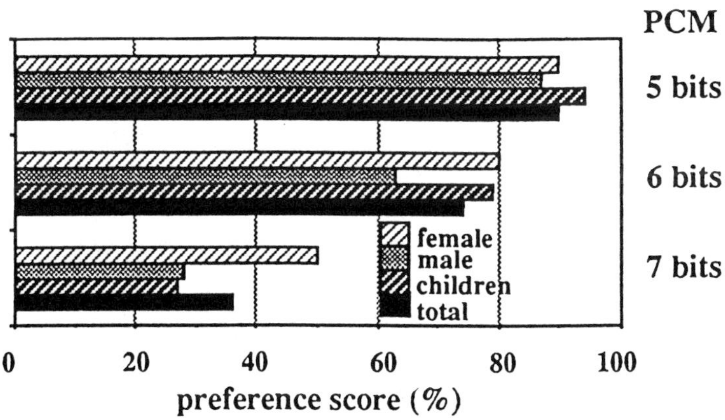

Fig.4 Results of pair-comparison tests

CONCLUSIONS

A low-delay high-quality 8-kbit/s speech coder has been designed. This coder is based on the combination of forward and backward prediction in the framework of a CELP coder. The frame length of 23 samples at 8 kHz sampling gives an algorithmic delay of 2.875 ms. Total coder delay will be three times as long as the algorithmic delay.

The proposed coder uses three novel schemes: a conditional pitch prediction scheme, backward adaptation of the gain and a switchover scheme for the synthesis filter. SNR of the coded speech is improved due to these schemes. Moreover, the SNR is significantly improved due to non-integer pitch delay and trained codebook. In total, the SNR of the proposed coder is 2 dB higher than that of the conventional backward adaptive CELP.

The quality of the proposed coder is noticeably superior to that of 6-bit PCM. Indeed, the quality of female speech is equivalent to that of 7-bit PCM. The proposed coder can give even higher quality if post-filtering is introduced.

For this coding scheme to be to applied to communication systems, the computational complexity should be reduced. Channel errors must also be investigated for cellular radio applications.

REFERENCES

[1] J. H. Chen and R. V. Cox: "A Fixed-Point 16kb/s LD_CELP Algorithm and Its Real-Time Implementation ," *Proc. ICASSP'91*, pp.21-24, 1991.

[2] M. Foodeei and P. Kabal: "Low-Delay CELP and Tree Coders: Comparison and Performance Improvements ," *Proc. ICASSP'91*, pp.25-28, 1991.

[3] J. Menez, C. Galand and M. Rosso: "A 2ms-Delay Adaptive Code Excited Linear Predictive Coder," *Proc. ICASSP'90*, pp.457-460, 1990.

[4] I. Gerson and M. Jasiuk: "Vector Sum Excited Linear Prediction (VSELP) Speech Coding at 8 kb/s", *Proc. ICASSP'90*, pp.461-464, 1990.

[5] T. Ohya, H. Suda, S. Uebayashi, T. Miki and T. Moriya: "Revised TC-WVQ Speech Coder for Mobile Communication System", *ICSLP'90* pp.125-128, 1990.

[6] N. S. Jayant: "High-Quality Coding of Telephone Speech and Wideband Audio," *IEEE Communications Magazine*, pp.10-20, Jan. 1990.

[7] V. Iyengar and P. Kabal: "A Low Delay 16 kb/s Speech Coder," *IEEE Tans.* SP-39(5), pp.1049-1057, May. 1991.

[8] M. R. Schröeder and B. S. Atal: "Code-Excited Linear Prediction (CELP): High-Quality Speech at Very Low Bit Rates", *Proc. ICASSP'85*, pp.937-940, 1985.

[9] P. Kroon and B. S. Atal: "Quantization Procedures for the Excitation in CELP coders," *Proc. ICASSP'87*, pp.1649-1652, 1987.

[10] N. S. Jayant and P. Noll: *Digital Coding of Waveforms,* Prentice-Hall, 1984.

[11] A. Kataoka and T. Moriya: "A Backward Adaptive 8kbit/s Speech Coder using Conditional Pitch Prediction", *GLOBECOM'91*, pp.1889-1893, 1991.

[12] J. H. Chen and A. Gersho: "Gain-Adaptive Vector Quantization with Application to Speech Coding," *IEEE Tans.* COM-35(9), pp.918-930, Sep. 1987.

[13] B. S. Atal and M. R. Schröeder: "Predictive Coding of Speech Signals and Subjective Criteria," *IEEE Tans.* ASSP-27(3), pp.247-254, Jun. 1979.

[14] S. P. Lloyd: "Least Squares Quantization in PCM," *IEEE Trans.* IT-28, pp.129-137, 1982.

[15] P. Kroon and B. S. Atal: "Pitch Predictors with High Temporal Resolution," *Proc. ICASSP'90*, pp.661-664, 1990.

3

LOW DELAY CODING OF SPEECH AND AUDIO USING NONUNIFORM BAND FILTER BANKS

Kambiz Nayebi and Thomas P. Barnwell

School of Electrical Engineering
Georgia Institute of Technology
Atlanta, GA 30332, U.S.A.

INTRODUCTION

Over the last decade, analysis-synthesis systems based on maximally decimated filter banks have emerged as one of the important techniques for speech and audio coding. For speech and audio signals, the analysis-synthesis filter bank can be thought of as modeling the human auditory system, where the critical band model of aural perception is reflected in the design of the filter banks. The constraints imposed by the aural model are best met by nonuniform analysis-synthesis systems in which the bandwidths of the channels increase with increasing frequency.

Tree-structured filter banks have been used to model the critical bands, but they fall short of a close approximation. In addition, tree-structured systems have the added disadvantage of inherent long reconstruction delays. Both of these problems can be addressed using a new reconstruction theory and design methodology which we have recently introduced [1, 2]. This theory results in a unified design methodology for all uniform and nonuniform analysis-synthesis systems based on FIR filter banks. This new approach for designing analysis-synthesis systems based on nonuniform band filter banks with arbitrary bandwidths [2, 3, 4] and low reconstruction delay [5, 6] has created many new possibilities for designing frequency domain audio and speech coders with very low reconstruction delays.

In this chapter, we present the design principles for the low delay and nonuniform filter banks, and we also present some details of a subband coder based on low delay, two-band systems. We show that the reconstruction delay of most existing subband coders can significantly be reduced without any noticeable degradation compared to the existing structures. This can simply be achieved by changing the analysis and the synthesis filters of the existing subband coders with the filters of the low delay systems.

LOW DELAY FILTER BANKS

All previously known analysis-synthesis filter banks with N-tap filters – such as those composed of quadrature mirror filters (QMF) and conjugate quadrature filters (CQF) – have $N-1$ samples of delay from the input to the output. A low delay filter bank system with N-tap filters has a reconstruction delay which is smaller than $N-1$. Designing such low and minimum delay systems is first achieved by the time-domain formulation of the system. In the time-domain formulation, the reconstruction conditions of the system are expressed in terms a matrix equation of the form

$$\mathbf{AS} = \mathbf{B} \tag{1}$$

where **A** contains the analysis filter coefficients and **S** contains the synthesis filter coefficients and matrix **B** is called the reconstruction matrix. In [1], we show that the structure of matrix **B** defines the reconstruction delay of the system. Assuming a maximally decimated uniform M-band system, matrix **B** is of the form

$$\mathbf{B} = [\mathbf{0}|\mathbf{0}|\ldots|\mathbf{J}_M|\mathbf{0}|\ldots|\mathbf{0}|\mathbf{0}]^T \tag{2}$$

where $\mathbf{0}$ is the $M \times 1$ zero vector, $\mathbf{J}_M$ is the $M \times M$ exchange matrix, and T denotes transposition. The position of J_M in matrix **B** determines the system delay. For example, in a critically sampled system, the minimum system delay is $M-1$ samples and is achieved when $\mathbf{J}_M$ is the first block of the **B** matrix and the maximum delay of $2N-M-1$ samples is obtained when $\mathbf{J}_M$ is the last block of **B**.

One design procedure based on the time-domain formulation is presented in [1]. In this procedure, a cost function containing the reconstruction error and frequency error is minimized to obtain proper filters with perfect or near perfect reconstruction. Another design approach is based on a constrained optimization procedure in which a frequency error is minimized subject to the reconstruction error being zero. Both methods have proven to be successful.

Two-Band Systems

Considering a two-band system with analysis filters $H_0(z)$ and $H_1(z)$, and synthesis filters $G_0(z)$ and $G_1(z)$, aliasing distortion is eliminated by choosing the synthesis filters as $G_0(z) = H_1(-z)$ and $G_1(z) = -H_0(-z)$ and the system transfer function can be expressed as $T(z) = F(z) + F(-z)$ where $F(z) = H_0(z)H_1(-z)$ is the product filter. For exact reconstruction $T(z)$ needs to be a pure delay, $z^{-\Delta}$, where Δ is the reconstruction delay of the system. This condition requires that every other sample of $f(n)$ (odd samples or even samples), except one sample, be equal to zero [6]. Any product filter that satisfies this condition can be decomposed into two filters $H_0(z)$ and $H_1(z)$ which result in a perfectly reconstructing system. Figure 1 shows the responses of the lowpass analysis filters of a two-band system with 8-tap and 16-tap system filters with 1 and 7 samples of delay respectively. Obviously, imposing a delay of $\Delta < N$ on a filter bank is a constraint that results in the reduced filter quality compared to the $\Delta = N$ case, and better quality as compared the system with shorter filters.

NONUNIFORM FILTER BANKS

Nonuniform filter banks in conjunction with tree-structures can be used to produce analysis-synthesis systems which can closely approximate critical

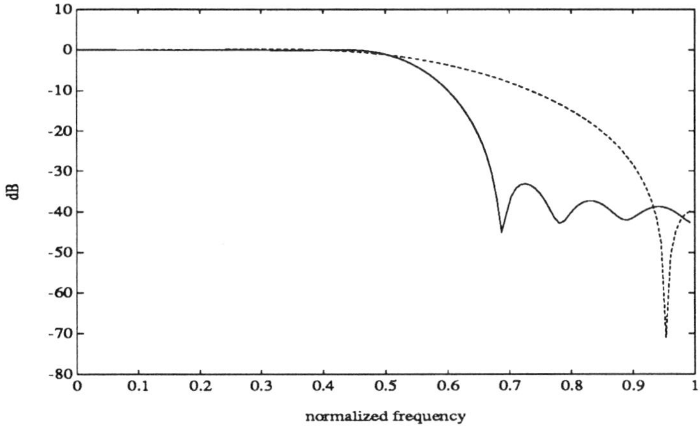

Figure 1: **The Lowpass Analysis Filters of Two-Band Systems with 16-Tap (Solid Line) and 8-Tap (Dashed Line) Filters with 7-samples and 1-sample Delays Respectively.**

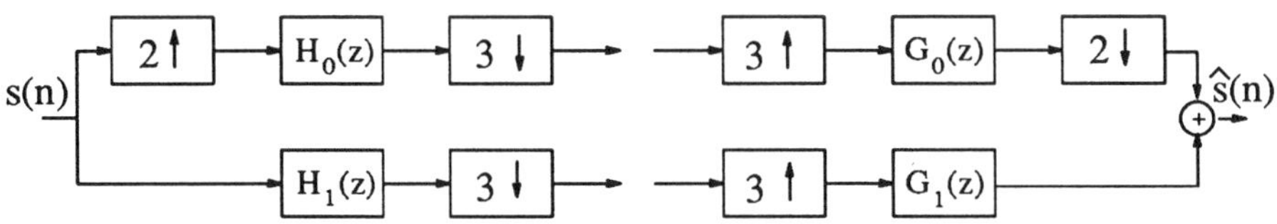

Figure 2: **A (2/3,1/3) Nonuniform Two-Band Filter Bank.**

bands. In this procedure, some two-band nonuniform systems are designed as basic splitting elements of the system. These two-band systems divide the signal spectrum into two unequal bands with different ratios. For example, in a two-band nonuniform system referred to as a $(\frac{p}{q}, 1-\frac{p}{q})$ system, the first band covers the frequency range $[0, \frac{p\pi}{q}]$, and the second band covers the $[\frac{p\pi}{q}, \pi]$ range. Figure 2 shows the block diagram of a (2/3,1/3) two-band nonuniform system. It is obvious that an M-band critical band system can be represented by a tree-structure of $M-1$ nonuniform two-band systems each with a proper ratio p_i/q_i. Most existing tree-structures are based on (1/2,1/2) systems with QMF's. Our recent experiments show that by using the combination of (2/3,1/3) and (1/2,1/2) systems, a significantly closer approximation of the critical bands can obtained. By designing more two-band nonuniform systems with different ratios, closer approximations of the critical bands are possible. Figure 3 shows the analysis filters of a (4/5,1/5) system.

Each of these nonuniform systems can also be designed to have a low reconstruction delay. Using the low delay systems will overcome the major disadvantage of the tree-structured systems which is the long system delay. Critical

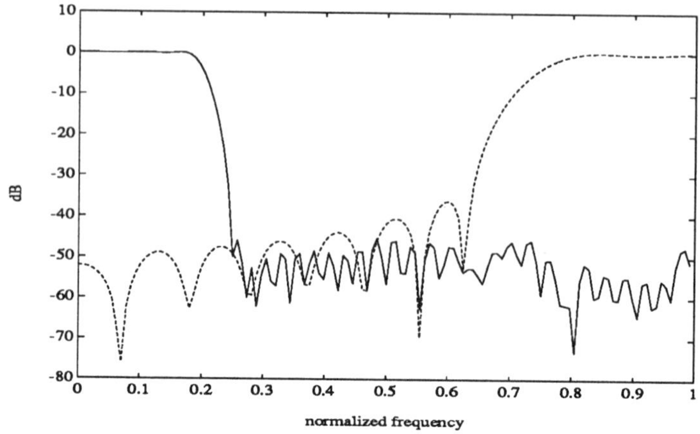

Figure 3: **Analysis Filters of the (4/5,1/5) System. The solid line is $H_0(z)$ and the dashed line is $H_1(z)$.**

band systems can also be designed using larger split and merge components with more than two-bands. For a 4 kHz speech signal, a total of about 15 to 25 critical bands are required for a close approximation of critical bands. For high quality 20 kHz music and audio signals, up to 40 critical bands are usually used.

LOW DELAY SUBBAND CODER

To show the effectiveness of low delay systems, we describe a speech subband coder which is based on a low delay tree-structured filter bank. Figure 4 is the tree-structure that was used in the coder which is based on low delay (1/2,1/2) split units. This coder is compared to another subband coder with a similar tree-structure which is based on QMFs. These systems use 32-tap, 16-tap, and 8-tap FIR filters. In the tree-structure under consideration, the speech is first split into four equal bands and the lowest band is further divided into four octave bands.

For the QMF system used in our experiments, this results in a system delay of 353 samples. This delay is reduced to 173 samples by using low delay two-band systems. In this case, systems with 32-tap, 16-tap, and 8-tap filters are designed with 15, 9, and 1 sample delays, respectively. The speech coder used as a basis for this test was a real-time implementation operating on a TI TMS320C31 floating point processor. A full duplex version of this coder uses about 20% of the processor's available cycles. Although the informal test we conducted cannot be said to be definitive, it showed that the low delay coder performs nearly as well as the QMF based coder for similar filter qualities at 16 Kbps. The cost is a higher number of multiplies and adds for the implementation of the low delay filter bank.

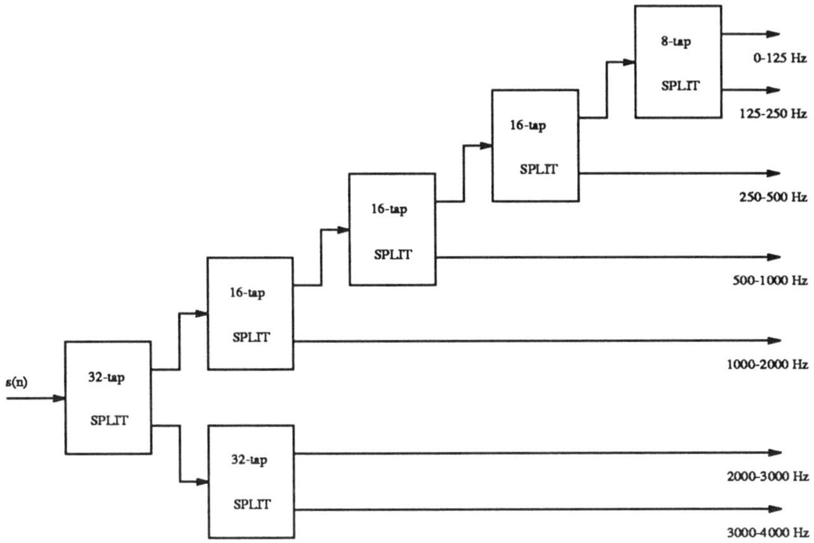

Figure 4: **The Tree-Structure of the Real-Time Subband Coder.**

References

[1] K. Nayebi, T. P. Barnwell, and M. J. T. Smith, "Time domain filter bank analysis: A new design theory," *IEEE Transactions on Signal Processing*, June 1992.

[2] K. Nayebi, T. P. Barnwell, and M. J. T. Smith, "The design of perfect reconstruction nonuniform band filter banks," *Proceedings of the International Conference on Acoustics, Speech, and Signal Processing*, pp. 1781–1784, 1991.

[3] J. Kovacevic and M. Vetterli, "Perfect reconstruction filter banks with rational sampling rate changes," *Proceedings of the International Conference on Acoustics, Speech, and Signal Processing*, pp. 1785–1788, 1991.

[4] P. Q. Hoang and P. P. Vaidyanathan, "Non-uniform mutirate filter banks: Theory and design," *Proceedings of the International Symposium on Circuits and Systems*, pp. 371–374, May 1989.

[5] K. Nayebi, T. P. Barnwell, and M. J. T. Smith, "Design of low delay FIR analysis-synthesis filter bank systems," *Proc. Conf. on Information Sciences and Systems*, pp. 233–238, 1991.

[6] K. Nayebi, T. P. Barnwell, and M. J. T. Smith, "Low delay FIR filter banks: Design and evaluation," *Accepted for publication in IEEE Trans. on Signal Processing.*

4

8 KB/S LOW-DELAY CELP CODING OF SPEECH

Juin-Hwey Chen and Martin S. Rauchwerk

AT&T Bell Laboratories
Murray Hill and Middletown, New Jersey, USA

INTRODUCTION

In the past few years, the CCITT's activities in standardizing a 16 kb/s low-delay speech coder has spurred significant research interest in low-delay speech coding. In response to this standardization effort, we have previously created a toll-quality 16 kb/s speech coder called Low-Delay CELP (LD-CELP) which has a one-way coding delay of less than 2 ms [1-6]. In May 1992, this 16 kb/s LD-CELP coder was officially adopted as the CCITT G.728 standard for 16 kb/s speech coding. Low-delay speech coding at 8 kb/s is the next natural target for research. Several researchers have worked in this area recently [7-14].

With our experience in 16 kb/s LD-CELP as a starting point, in 1989 we started out to explore the possibility of low-delay CELP speech coding at 8 kb/s. Our goal was to match the speech quality of conventional 8 kb/s CELP coders under (1) a one-way delay constraint of around 10 ms, (2) a complexity constraint that a full-duplex coder should fit on a single DSP, and (3) a robustness constraint that two-way conversation should not have difficulties at a bit-error rate (BER) up to 10^{-3}. Bounded by these stringent constraints, we found it to be a formidable task to achieve our goal. In this paper, we will describe our 8 kb/s LD-CELP coder, its real-time implementation, and its performance.

SYSTEM OVERVIEW

Although we started with the 16 kb/s LD-CELP structure, we had to make several major changes to achieve good speech quality at 8 kb/s. Figure 1 shows the resulting 8 kb/s LD-CELP encoder. Due to the delay constraint, backward adaptation was still used to update a 10th-order LPC predictor and the excitation gain. The pitch parameters, however, were forward transmitted to achieve higher speech quality and better robustness to channel errors. We designed a 3-tap pitch predictor where the pitch period was inter-frame differentially coded into 4 bits and the 3 tap-weights were vector quantized to 5 or 6 bits, with the codebook search jointly optimizing the pitch period and the 3 taps in a closed-loop manner. The decoder (not shown here) used an adaptive postfilter similar to the one proposed in [15] and [16].

LPC PREDICTION

The one-way coding delay of a CELP coder is typically 2.5 to 3 times its frame buffer size. Thus, our 10 ms delay constraint limited the maximum frame size to 4 ms. To investigate the trade-off between coding delay and speech quality, we created two coder versions with different frame sizes: 2.5 ms and 4 ms. At 8 kb/s and with 8 kHz sampling, this gave us only 20 or 32 bits to spend in each frame — clearly not enough for forward transmission of both the LPC parameters and the excitation. Thus, it was necessary to make the LPC predictor backward-adaptive.

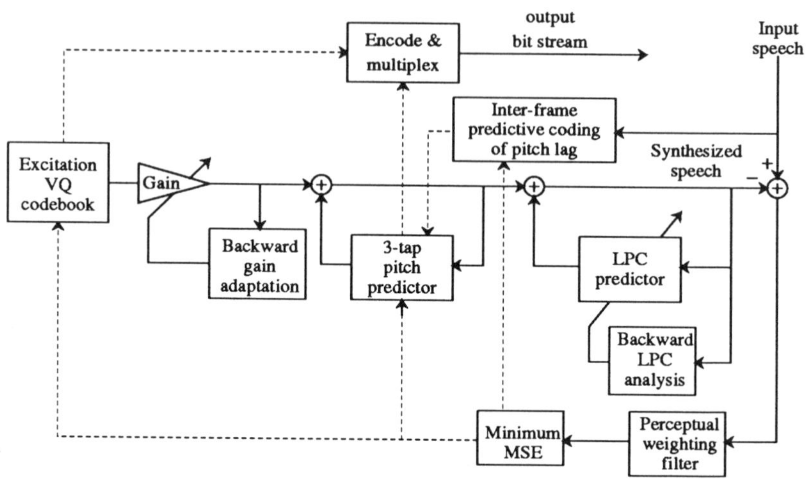

Fig. 1 Block diagram of the 8 kb/s LD-CELP coder

Starting with 16 kb/s LD-CELP, we first doubled the excitation vector dimension from 5 to 10 samples and otherwise kept the algorithm the same. The resulting 8 kb/s LD-CELP coder produced rather noisy speech. To investigate the problem, we first reduced the LPC predictor order from 50 back to 10. The resulting speech was about as noisy as before. This indicated that the 50th-order backward-adaptive LPC predictor was not effective for 8 kb/s LD-CELP, although in 16 kb/s LD-CELP it was proven useful for exploiting the pitch redundancy. In our next diagnostic experiment, we performed the LPC analysis on previous input speech rather than coded speech. The resulting speech had a lower coding noise, as expected, but the speech quality was still not satisfactory. The conclusion from these experiments was clear: to achieve good speech quality at 8 kb/s with low delay, we needed to exploit the pitch redundancy explicitly. Hence, we added a pitch predictor and reduced the LPC predictor order to 10. The 10th-order LPC predictor was backward adapted once a frame using the autocorrelation method of LPC analysis.

PITCH PREDICTION

Initially, we tried a backward-adaptive 3-tap pitch predictor [17] with resets during unvoiced or silent frames [18]. However, even with frequent resets, the robustness to channel errors was still not satisfactory at BER = 10^{-3}. We also tried backward adaptation for either the pitch period or the predictor taps and forward transmit the other, but such schemes were still sensitive to channel errors.

The next possibility was the approach proposed in [7], where the predictor tap was forward transmitted but the pitch period was partially backward and partially forward adapted. The forward adaptation part of the pitch period was completely based on the result of backward adaptation, which was known to be sensitive to channel errors. Hence, the entire scheme is expected to be sensitive to errors as well, so we did not try this scheme. This left us with the only choice: fully forward-adaptive pitch prediction.

We first developed a 3-tap pitch predictor with the pitch period closed-loop quantized to 7 bits and the 3 taps closed-loop vector quantized to 5 or 6 bits. This scheme achieved a high pitch prediction gain and was much more robust to channel errors than any of the pitch predictor schemes we tried above. However, with only 20 or 32 bits available for each frame, spending 12 or 13 bits on the pitch predictor left too few bits for excitation coding, especially in the 2.5 ms frame version. Thus, the pitch predictor encoding rate had to be reduced.

Initially, we tried to change to a single-tap pitch predictor so that only 3 bits were needed to specify the tap. Unfortunately, this resulted in noticeable degradation in speech quality. Hence, we were forced to try the alternative — reducing the encoding rate of the pitch period. Since pitch periods in adjacent frames were highly correlated, inter-frame predictive coding could be used to reduce the encoding rate of the pitch period. The challenges, however, were: (1) how to make the scheme robust to channel errors, (2) how to track the sudden change in the pitch period quickly at the beginning of each voiced region, and (3) how to maintain the high prediction gain in voiced regions. With these challenges in mind, we designed a 4-bit inter-frame predictive coding scheme for the pitch period.

To enhance the robustness to channel errors, we used a simple first-order, fixed-coefficient predictor for inter-frame prediction. We made the predictor "leaky" so that channel error effects would decay with time. Another way we improved the robustness was by using pseudo Gray coding [19-21] on the codebook of the 3 pitch predictor taps. In addition, whenever the current frame was not voiced, we turned off the 3-tap pitch predictor and reset the inter-frame predictive coding scheme. This further confined channel error effects to within one voiced segment. Note that the decoder needed to be signaled when to turn off the pitch predictor and perform the reset. Rather than sending a dedicated bit once a frame to indicate whether the current frame was voiced, we did it in a more efficient way. We "stole" one quantizer level from the 4-bit prediction-error quantizer. During voiced frames, only 15 out of the 16 possible quantizer levels were used to quantize the inter-frame prediction error of the pitch period. The 16th quantizer level was sent to the decoder only if the current frame was not voiced, and thus could be used as a voicing flag. Such "quantizer level stealing" avoided the need to spend an extra bit on the voicing flag, and it did not cause noticeable degradation in the quantizer's performance. To reduce the probability that a bit error in the 4-bit quantizer index caused the decoder to turn off the pitch predictor erroneously, we also sent a special all-zero codevector from the pitch tap codebook (i.e. all 3 pitch taps were zero) whenever the current frame is not voiced. The decoder would turn off the pitch predictor and perform the reset only if it received both the all-zero pitch tap codevector and the 16th level of the 4-bit quantizer.

To meet the challenge of quickly tracking the sudden change in the pitch period at speech onsets, we used non-uniformly spaced quantizer levels in the 4-bit quantizer. The outer quantizer levels were large enough to quickly catch up with the sudden pitch change within 2 to 3 frames (5 to 12 ms) after speech onsets. The closely spaced inner quantizer levels were close enough to track the subsequent slow pitch changes with the same precision as the conventional 7-bit instantaneous pitch quantizer. This helped to maintain the high prediction gain in voiced regions.

An additional step in meeting the challenge of maintaining a high prediction gain was to perform closed-loop joint quantization of the pitch period and the 3 pitch predictor taps. We generalized the closed-loop quantization procedure for a single-tap pitch predictor [22] to the 3-tap case. Ideally, for best performance, we should search through all combinations of the pitch quantizer levels and the codevectors of the 3-tap VQ codebook. However, this would exceed our real-time budget in DSP implementation. To reduce the complexity, we allowed only a subset of pitch quantizer levels while performing the closed-loop search through the 3-tap codebook.

Our pitch parameter coding scheme described above achieved roughly the same pitch prediction gain (5 to 6 dB) as our initial scheme with the 7-bit pitch period and 5 or 6-bit pitch taps. Furthermore, with noisy channels, comparable speech quality was obtained whether we used the 7-bit pitch quantizer or our 4-bit inter-frame predictive quantizer. Thus, we have reduced the pitch period encoding rate from 7 bits/frame to 4 bits/frame without compromising the pitch prediction gain or the robustness to channel errors. Saving 3 bits per frame may appear insignificant, but with our small frame sizes, such saving accounts for 10 to 15% of the total bit-rate, or 750 to 1200 bps. Allocating these 3 bits to excitation coding improved speech quality significantly.

GAIN ADAPTATION

The gain adaptation scheme is essentially the same as in the 16 kb/s LD-CELP algorithm. The excitation gain is backward-adapted by a 10th-order linear predictor operated in the logarithmic gain domain. The coefficients of this gain predictor are updated once a frame by backward-adaptive LPC analysis on previous logarithmic gains of scaled excitation vectors [2].

EXCITATION CODING

The 2.5 ms frame version has one excitation vector per frame, while the 4 ms frame version has two 16-dimensional excitation vectors per frame. Similar to 16 kb/s LD-CELP, we also used a gain-shape structured excitation codebook to reduce the codebook search complexity. The codebook was designed by a closed-loop training procedure [2]. The codebook indices were pseudo Gray coded to improve the robustness to channel errors. Table 1 summarizes the coder parameters and bit allocation of the two versions of 8 kb/s LD-CELP.

Coder version number	1	2
Frame size (ms)	2.5	4
Frame size (samples)	20	32
Vector dimension	20	16
Vectors/frame	1	2
Pitch period (bits)	4	4
Pitch taps (bits)	5	6
Excitation sign (bit)	1	1×2
Excitation magnitude (bits)	3	3×2
Excitation shape (bits)	7	7×2
Total number of bits/frame	20	32

Table 1 Coder parameters and bit allocation of 8 kb/s LD-CELP

REAL-TIME IMPLEMENTATION

We have implemented a real-time, full-duplex 8 kb/s LD-CELP coder using a single 80 ns AT&T DSP32C chip. We chose to implement the 4 ms frame version, because it produced better speech quality with a lower complexity. Table 2 shows the processor time and memory usage of this implementation. The encoder and decoder take 80.1% and 12.4% of the processor time, respectively. A full-duplex coder requires 40.91 kbytes (or about 10 kwords) of memory. This includes the 1.5 kwords of RAM on the DSP32C chip.

Implementation mode	Processor time (% DSP32C)	Program ROM (kbytes)	Data ROM (kbytes)	Data RAM (kbytes)	Total memory (kbytes)
Encoder only	80.1%	8.44	20.09	6.77	35.29
Decoder only	12.4%	3.34	11.03	3.49	17.86
Full-duplex	92.5%	10.50	20.28	10.12	40.91

Table 2 DSP32C processor time and memory usage of 8 kb/s LD-CELP

PERFORMANCE

In a formal subjective listening test, the 13 kb/s GSM coder (European cellular standard) [23], the 8 kb/s VSELP coder (North American cellular standard) [24], and the 4 ms frame version of the 8 kb/s LD-CELP coder all achieved almost identical Mean Opinion Scores. The 2.5 ms frame version scored very close to the 4 ms frame version — only 0.04 lower in MOS. Therefore, if certain applications demand a one-way delay of 7 ms or less, this 2.5 ms version can be used. In a more recent test, an improved 8 kb/s LD-CELP coder (4 ms frame) achieved an MOS 0.09 higher than than the MOS of the 8 kb/s VSELP coder obtained in the same test. The 8 kb/s

VSELP coder is a variant of conventional CELP with a frame size of 20 ms. Thus, 8 kb/s LD-CELP achieved a slightly higher MOS with only 1/5 of the delay.

The 8 kb/s LD-CELP coder is reasonably robust to channel errors without error protection. Although we noticed some quality degradation, we could communicate without difficulties in two-way telephone conversation when we talked through our real-time coders and a real-time simulated noisy channel with BER = 10^{-3}.

CONCLUSION

We have described our work in 8 kb/s Low-Delay CELP coding of speech. The main features of our algorithm are the backward-adaptive LPC predictor and excitation gain, and the 3-tap pitch predictor with inter-frame predictively coded pitch period and closed-loop vector quantized predictor taps. The main contribution of this work is to find a new way of CELP speech coding at 8 kb/s which, when compared with conventional CELP at the same bit-rate, achieves the same or slightly better speech quality with roughly the same complexity but a much lower coding delay.

References

1. J.-H. Chen, "A robust low-delay CELP speech coder at 16 kbit/s," *Proc. IEEE Global Commun. Conf.*, pp. 1237-1241 (November 1989).
2. J.-H. Chen, "High-quality 16 kb/s speech coding with a one-way delay less than 2 ms," *Proc. IEEE Int. Conf. Acoust., Speech, Signal Processing,,* pp. 453-456 (April 1990).
3. J.-H. Chen, M. J. Melchner, R. V. Cox, and D. O. Bowker, "Real-time implementation and performance of a 16 kb/s low-delay CELP speech coder," *Proc. IEEE Int. Conf. Acoust., Speech, Signal Processing,,* pp. 181-184 (April 1990).
4. J.-H. Chen, Y.-C. Lin, and R. V. Cox, "A Fixed-Point 16 kb/s LD-CELP Algorithm," *Proc. IEEE Int. Conf. Acoust., Speech, Signal Processing,,* pp. 21-24 (May 1991).
5. J.-H. Chen, N. S. Jayant, and R. V. Cox, "Improving the performance of the 16 kb/s LD-CELP speech coder," *Proc. IEEE Int. Conf. Acoust., Speech, Signal Processing,,* pp. I-69 to I-72 (March 1992).
6. J.-H. Chen, R. V. Cox, Y.-C. Lin, N. S. Jayant, and M. J. Melchner, "A low-delay CELP coder for the CCITT 16 kb/s speech coding standard," *IEEE J. Selected Areas Communications*, pp. 830-849 (June, 1992).
7. T. Moriya, "Medium-delay 8 kbit/s speech coder based on conditional pitch prediction," *Proc. of Int. Conf. Spoken Language Processing*, (November 1990).
8. J.-H. Chen and M. S. Rauchwerk , "An 8 kb/s low-delay CELP speech coder ," *Proc. IEEE Global Comm. Conf.*, pp. 1894-1898 (December 1991).
9. J. D. Gibson and H. Woo, "Low delay tree coding of speech at 8 kbps," *Proc. IEEE Global Comm. Conf.*, pp. 1884-1888 (December 1991).

10. A. Kataoka and T. Moriya, "A backward adaptive 8 kbit/s speech coder using conditional pitch," *Proc. IEEE Global Comm. Conf.*, pp. 1889-1893 (December 1991).

11. S. Ono, "8 kbps low delay celp with feedback vector quantization," *Proc. IEEE Global Comm. Conf.*, pp. 700-704 (December 1991).

12. J.-H. Yao, J. Shynk, and A. Gersho, "Low-delay vector excitation coding of speech at 8 kbps," *Proc. IEEE Global Comm. Conf.*, pp. 695-699 (December 1991).

13. R. Soheili, A. Kondos, and B. Evans, "Techniques for improving the quality of LD-CELP coders at 8 kb/s," *Proc. IEEE Int. Conf. Acoust., Speech, Signal Processing,*, pp. I-41 to I-44 (March 1992).

14. J.-H. Yao, J. Shynk, and A. Gersho, "Low-delay VXC at 8 kb/s with interframe coding," *Proc. IEEE Int. Conf. Acoust., Speech, Signal Processing,*, pp. I-45 to I-48 (March 1992).

15. J.-H. Chen, *Low-bit-rate predictive coding of speech waveforms based on vector quantization,* Ph. D. dissertation, University of California, Santa Barbara (March 1987).

16. J.-H. Chen and A. Gersho, "Real-time vector APC speech coding at 4800 bps with adaptive postfiltering," *Proc. IEEE Int. Conf. Acoust., Speech, Signal Processing*, pp. 2185-2188 (April 1987).

17. V. Iyengar and P. Kabal, "A low delay 16 kbits/sec speech coder," *Proc. IEEE Int. Conf. Acoust., Speech, Signal Processing*, pp. 243-246 (April 1988).

18. R. Pettigrew and V. Cuperman , "Backward pitch prediction for low delay speech coding," *Proc. IEEE Global Comm. Conf.*, pp. 1247-1252 (November 1989).

19. J.R.B. De Marca and N.S. Jayant, "An algorithm for assigning binary indices to the codevectors of a multi-dimensional quantizer," *Proc. IEEE Int. Conf. on Communications*, pp. 1128-1132 (June 1987).

20. K.A. Zeger and A. Gersho, "Zero redundancy channel coding in vector quantization," *Electronics Letters* **23**(12) pp. 654-656 (June 1987).

21. K. Zeger and A. Gersho, "Pseudo-Gray coding," *IEEE. Trans. Communications*, pp. 2147-2158 (December 1990).

22. W.B. Kleijn, D.J. Krasinski, and R.H. Ketchum, "Improved speech quality and efficient vector quantization in SELP," *Proc. IEEE Int. Conf. Acoust., Speech, Signal Processing*, (April 1988).

23. P. Vary et al., "Speech codec for the European mobile radio system," *Proc. IEEE Global Comm. Conf.*, (November 1989).

24. I. Gerson and M. A. Jasiuk, "Vector sum excited linear prediction (VSELP) speech coding at 8 kbps," *Proc. IEEE Int. Conf. Acoust., Speech, Signal Processing,*, pp. 461-464 (April 1990).

5

LATTICE LOW DELAY VECTOR EXCITATION FOR 8 kb/s SPEECH CODING

Aamir Husain and Vladimir Cuperman

Communication Sciences Laboratory, School of Engineering Science,
Simon Fraser University, Burnaby, B.C V5A 1S6, Canada

INTRODUCTION

Recently, communication delay has become an important performance criterion for speech encoders used in the public switched telephone network (PSTN), as efforts are being intensified to achieve toll quality at rates as low as 8 kb/s, to replace existing higher rate systems. In a complex network, the delays of many encoders add together, transforming the delay into a significant impairment of the system. Delay may necessitate the use of echo cancellation and in some applications it remains an impairment even after echo cancellation has been performed. For these reasons, the proposed 8 kb/s CCITT standard specifies low delay as a major requirement.

The delay performance of a speech coder is characterized by the algorithmic delay and processing delay. Algorithmic delay is the one-way delay of the encoder and the decoder assuming infinite processing power for the coder implementation. Processing delay is the additional delay due to implementation with a finite processing power. The total codec delay is the sum of algorithmic delay and processing delay. The channel delay caused by transmitting over a finite bandwidth serial channel adds to the total codec delay to give the delay encountered in practical applications. The requirements of the new 8 kb/s CCITT standard specify a frame size lower than 16 ms (objective 5 ms) and a total codec delay of 32 ms (objective 10 ms).

Conventional Code Excited Linear Prediction (CELP) [1] or Vector Excitation Coding (VXC) [2] achieve good speech quality; however these coders introduce a substantial delay due to forward adaptation of the short-term predictor. The input buffering delay, typically 20 ms at 8 kHz and other processing delays, typically result in a total codec delay of 50 to 60 ms. Although a total delay of 32 ms may be obtained by re-designing the standard CELP configuration for a different delay/quality trade-off, lower delays (which can meet or exceed the objective of 10 ms total codec delay) and good quality can be obtained by using a backward adaptive configuration.

In a backward adaptive analysis-by-synthesis configuration, the parameters of the synthesis filter are not derived from the original speech signal, but instead computed by backward adaptation, extracting information only from the reconstructed signal based on the transmitted excitation information. Since both the encoder and decoder have access to the past reconstructed signal, side information is no longer needed for the synthesis filter, and the low-delay requirement can be met with a suitable choice of frame size.

Backward adaptive analysis-by-synthesis configurations are used for speech coding at 16 kb/s in [3-6]. The recently adopted G.728 16 kb/s CCITT speech coding standard is based on Low-Delay CELP and achieves toll quality with a delay lower than 2 ms using a block backward adaptive configuration without pitch prediction [6]. The Lattice Low-Delay VXC (LLD-VXC) achieves similar performance using backward adaptive lattice adaptation and backward adaptive pitch prediction [5].

Recently, several low delay 8 kb/s coders have been proposed which achieve close to toll quality. These include versions of LD-CELP and LD-VXC [7-10], and a tree encoder based on backward adaptive prediction [11]. Significant progress in improving backward adaptive configuration robustness on noisy channels was recently achieved [11].

This paper presents two versions of the 8 kbits/s LLD-VXC codec: a backward 8 kb/s coder, which makes use of a 3-tap hybrid backward adaptive open-loop pitch predictor [12, 13] and a partially-forward scheme, which uses a 3-tap forward adapted long-term adaptive codebook. These two codecs are compared in clean and noisy channel conditions.

SYSTEM OVERVIEW

In the block diagram of the backward LLD-VXC system shown in Fig. 1, each candidate excitation codevector **c** is multiplied by a gain, and the resulting vector, **u**, is fed into the synthesis filters. The gain is the product of the predicted gain obtained from the backward adaptive gain predictor and the gain value obtained from the gain codebook. The output of the synthesis filters, **y**, is then compared to the actual speech signal, **x**, and the best candidate codevector is selected using a perceptually weighted minimum mean-square error criterion.

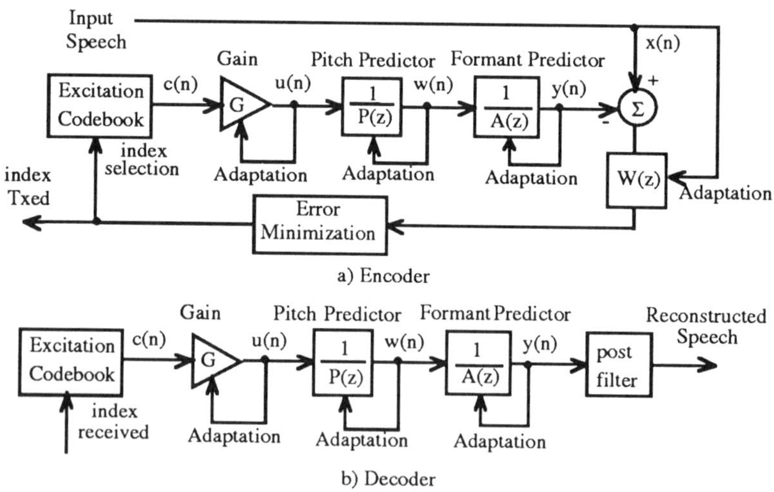

Figure 1: Backward Lattice LD-VXC Configuration

The index of the optimal excitation sequence is then transmitted to the decoder. At the decoder side, the received indices are used to generate the proper excitation sequence.The excitation codevector is then gain scaled using the gain computed in the same way as it is done in the encoder, and fed into the cascade of the pitch and formant synthesis filters.

The block diagram of the partially-forward LLD-VXC system is shown in Fig. 2. In the partially-forward system an additional gain codebook is used, which consists of the tap gains of the 3-tap long-term adaptive codebook.

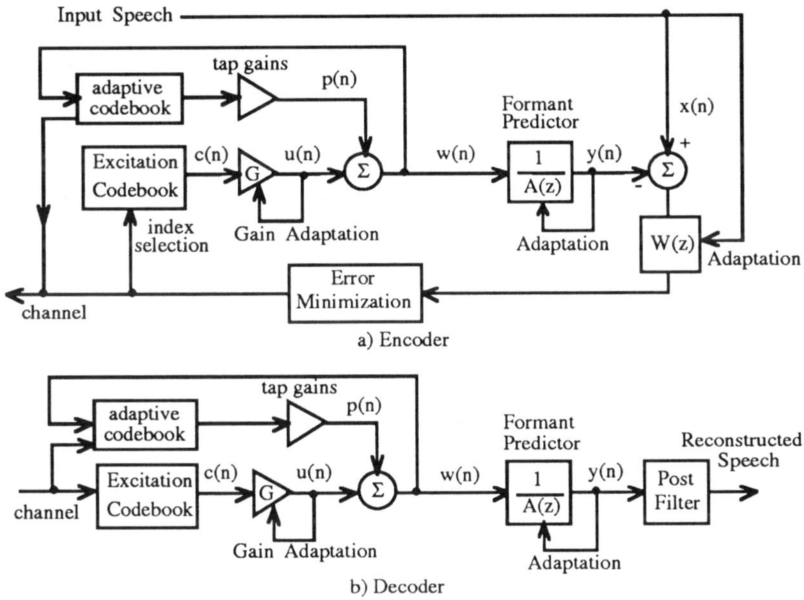

Figure 2: Partially-Forward Lattice LD-VXC Configuration

In the diagram shown in Fig. 2, each candidate excitation codevector **c** is multiplied by a gain, the resulting vector, **u**, is then added to the output vector of the adaptive codebook, **p**, to give the excitation vector, **w**, which is fed into both the short-term synthesis filter and the adaptive codebook memory. The adaptive codebook output, **p**, is obtained by a closed-loop codebook search procedure which selects the best delay and tap gains. The adaptive codebook search is discussed later in this paper. The output of the synthesis filter, **y**, is then compared to the actual speech signal, **x**, and the best candidate codevector is selected using a perceptually weighted minimum mean square error criterion based on the weighting filter, W(z).

The optimal excitation shape and gain indices as well as the adaptive codebook delay and tap gains indices are then transmitted to the decoder. At the decoder side, the received indices are used to generate the proper excitation sequence. The gain scaled excitation codevector is then added to the output of the adaptive codebook, the resultant being fed into both the formant synthesis filter and the adaptive codebook memory.

The 8 kb/s LLD-VXC system makes use of a 10th order perceptual weighting filter identical to that used in the 16 kb/s LLD-VXC system [5]. The excitation gain adaptation scheme is essentially the same as the 16 kb/s LLD-VXC algorithm [5]. The fixed prediction coefficients, p_i have been optimized on a large training set for each vector dimension investigated [10].

SHORT TERM PREDICTOR ADAPTATION

The 8 kb/s system makes use of a lattice structure for short-term prediction hence it is referred to as Lattice LD-VXC (LLD-VXC). For the 8 kb/s LLD-VXC system, the short-term predictor adaptation is essentially the same as for the 16 kb/s LLD-VXC algorithm [5, 10]. The adaptation is based on the least mean square (LMS) algorithm, with a leakage factor introduced to improve performance in noisy channel conditions [5]. The leakage and exponential weighting factors are optimized so as to achieve robustness in noisy channel conditions without a noticeable degradation in quality for clean channel conditions.

In the 16 kb/s LLD-VXC system the driving signal for coefficient adaptation was the reconstructed speech signal. In the 8 kb/s system, various driving signals were investigated to examine the effect of coefficient adaptation on system performance in clean and noisy channel conditions. The signals used are defined below (Fig. 3).

- y(n) - reconstructed speech signal.
- u(n) - excitation signal.
- $u_s(n)$ - u(n) passed through a short-term shaping filter.
- $u_l(n)$ - u(n) passed through a long-term shaping filter.
- $u_{ls}(n)$ - u(n) passed through long and short-term shaping filters.

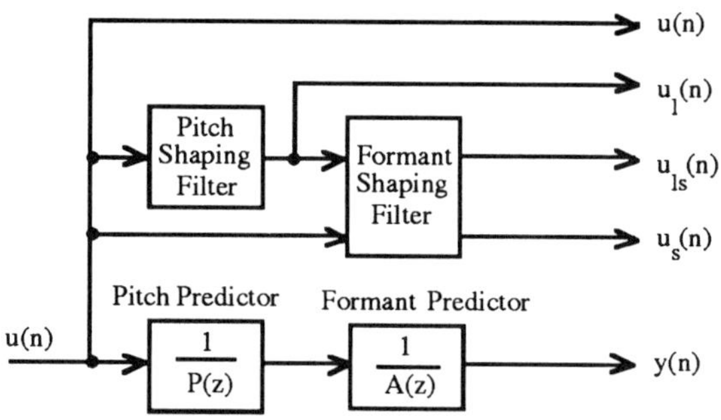

Figure 3: Lattice Filter adaptation signals

The short-term shaping filter is an FIR approximation of the short-term synthesis filter and is similar to the one described by Woo and Gibson in [11]. The long-term shaping filter, $P_{fir}(z)$, is obtained as a truncated finite impulse response approximation of the long-term predictor. Denote by P(z) the equivalent transfer function for the long-term predictor. Then, for a one-tap predictor, $P_{fir}(z)$ is given by

$$\frac{1}{1-P(z)} \approx 1 + P_{fir}(z) \qquad (1)$$

$$P_{fir}(z) = \sum_{i=1}^{M} [P(z)]^i = \sum_{i=1}^{M} [bz^{-k_p}]^i \tag{2}$$

M is the number of non-zero terms in $P_{fir}(z)$ for a one-tap predictor. Equation (2) can be easily extended for a three-tap case. Simulation results are shown in Table 1 for the backward system using various adaptation driving signals shown in Fig. 3.

Table 1 Seg SNR results for the backward system for various adaptation signals

Signal	$y(n)$	$u(n)$	$u_l(n)$	$u_s(n)$	$u_{ls}(n)$
BER=0	12.95	12.12	12.53	12.72	12.94
BER=10^{-3}	9.47	10.06	10.44	10.55	10.66

The results in Table 1 show that using FIR approximations of the short- and long-term synthesis filters in the adaptation loop improves performance on noisy channels without any significant degradation in clean channel conditions. Particularly, the adaptation signal, $u_{ls}(n)$, achieves comparable performance to y(n) in clean conditions and out-performs all other adaptation signals in noisy conditions. The backward system uses $u_{ls}(n)$ for adaptation of the short-term synthesis filter.

In the partially-forward system, $u_{ls}(n)$ provides a small performance improvement in noisy channel conditions. However, this improvement comes at the expense of a degradation in clean conditions which is roughly equivalent to the improvement in noisy conditions. As a result, the partially-forward system employs y(n) for the adaptation of the short-term synthesis filter.

ADAPTIVE CODEBOOK SEARCH AND PARAMETER ENCODING

The partially-forward LLD-VXC system makes use of an adaptive codebook similar to that used in forward CELP. The drawback of forward adaptation is the necessity to increase the vector dimension of the coder to accommodate the extra bits required for adaptive codebook index information.

A four-bit delta pitch encoder [7, 8, 10] is used thereby saving 3 bits per vector when compared to a conventional 7-bit pitch quantizer. The delta pitch encoder needs to have highly correlated pitch values in adjacent frames. This is difficult to achieve in a closed-loop search environment, as pitch doubling and tripling is very common. To solve this problem, an open-loop estimate of the pitch is obtained. This together with the previous frame's pitch is used in making the decision whether to use a delta pitch encoder (locked mode) or the conventional pitch quantizer (unlocked mode) for the current frame [7, 8, 10]. The delay of the adaptive codebook is referred to as the pitch for simplicity though it may not necessarily be an estimate of the pitch.

The 3 taps of the adaptive codebook are vector quantized to 6 bits (unlocked) or 7 bits (locked) using a tap-gain VQ. The first step in the VQ search procedure is selecting a set of possible candidate pitch values. A search is then performed for obtaining the best possible entry in the tap-gain VQ for each pitch candidate previously selected. Finally, the best pitch candidate and tap-gain VQ index that minimize the distortion measure are selected. For the locked mode, the delta pitch encoder is used, and the pitch candidates are selected from 16 possible pitch values, by performing a closed loop search using the optimal gains. However, for the unlocked mode, the conventional 7-bit pitch quantizer is used and the pitch candidates are selected from 128 possible pitch values.

CODEBOOK DESIGN

The optimization procedure of the excitation shape-gain codebook is the same as that of the 16 kb/s LLD-VXC system [5]. The tap gains codebook design problem for the adaptive codebook, can be defined as follows: given the training sequence of input speech vectors $\{X_n;\ n = 1, \ldots, N\}$ and the *ith* tap gains codebook of size P, find the *(i+1)th* "new" tap gains codebook which will minimize the *ith* average distortion

$$D = \sum_{n=1}^{N} \| X_n^* - \mathbf{Z}_n g_p \|^2 \tag{3}$$

where

$$\mathbf{Z}_n = \left[Z_{n-1}\ Z_n\ Z_{n+1} \right] \tag{4}$$

g_p is the *pth* entry of the tap gains codebook and vector $\mathbf{Z}_n$ is the unscaled adaptive excitation passed through the zero-state short-term filter. The vector X_n^* is the input speech vector, X_n, minus the zero input response of the short-term filter. For simplification, index n has been suppressed for g_p. By making use of variational techniques, the centroid of the tap gains codebook can be obtained by

$$\left(\sum_{n = G^p} Z_n^T Z_n \right) g_p^{new} = \sum_{n = G^p} Z_n^T X_n^* \tag{5}$$

where G^p is the *pth* cluster of the target vector, with g_p^{new} as the corresponding gain centroid of the adaptive codebook.

SIMULATION RESULTS

At 16 kb/s, in the presence of a pitch predictor, the short-term prediction gain saturates for a 20th order predictor for male and female speakers [4, 5]. At 8 kb/s there is no performance improvement for predictors of order larger than 10 [5]. The poor performance of high-order predictors at 8 kb/s may be caused by the quantization noise present in the adaptation loop.

In order to compare the backward open-loop system with the partially-forward closed-loop system, the following LLD-VXC systems were tested:

- System 1 - backward system, frame size 10, 8 bits shape and 2 bits gain codebooks
- System 2 - partially-forward system, frame size 24, bit allocation as shown in Table 2

Table 2 Bit allocation for partially-forward system

bit allocation	flag	shape cbk	gain cdbk	pitch	tap gains
locked	1	8	4	4	7
unlocked	1	8	2	7	6

Both the above systems use a 10th order lattice short-term predictor and a fixed 10th order gain predictor, optimized for each vector dimension. It was found experimentally that, even though the coefficient optimization of the gain predictor did not provide any significant objective performance improvement, it did offer marginal perceptual improvement. The results of simulation tests are shown in Table 3 below.

Table 3 Seg SNR results for the backward and partially-forward systems

System	BER=0	BER=10^{-3}
1	12.94	10.66
2	14.09	6.10

Table 3 shows that the partially-forward system has better performance in clean channel conditions, while the backward system has better performance on a noisy channel. However, informal subjective listening tests show that the forward system performance at a BER of 10^{-3} can be significantly improved by post-filtering. With post-filtering, the subjective performance of the partially-forward system is roughly equivalent to the performance of the backward system.

Informal subjective tests indicate that the backward and partially-forward 8 kb/s systems have quality comparable to the 8 kb/s VSELP standard in clean conditions. For noisy channels, at bit error rates of 10^{-3}, both systems achieve MOS scores which are within 0.2 on the MOS scale from the scores obtained in clean conditions. In the backward system, the use of the short-term adaptation signal $u_{ls}(n)$, results in a robust codec, which achieves good subjective quality, even at bit error rates as high as 10^{-2}. The partially-forward system degrades significantly at 10^{-2}, mainly due to the errors on the forward transmitted tap gains.

CONCLUSION

Informal MOS tests indicate that both the backward and partially-forward systems achieve subjective speech quality comparable to the 8 kb/s VSELP speech coder for the North American digital cellular system. Both systems also achieve good performance in noisy channel conditions and are therefore quite robust in the presence of channel errors.

REFERENCES

[1] B. S. Atal and M. R. Schroeder, "Stochastic Coding of Speech at Very Low Bit Rates", Proc. IEEE Int. Comm. Conf., 1984, pp. 1610-1613.

[2] G. Davidson, M. Yong, and A. Gersho, "Real-Time Vector Excitation Coding of Speech at 4800 BPS", Proc. IEEE ICASSP, Apr. 1987, pp. 2189-2192.

[3] L. Watts and V. Cuperman, "A Vector ADPCM Analysis-by-Synthesis Configuration for 16 kb/s Speech Coding", Proc. IEEE Globecom 1988, Conf., pp. 275-279.

[4] V. Cuperman, A. Gersho, R. Pettigrew, J. J. Shynk and J-H Yao, "Backward Adaptation for Low Delay Vector Excitation Coding of Speech at 16 kb/s", Proc. IEEE Globecom 1989, pp. 34.2.1-34.2.5.

[5] R. Peng, V. Cuperman, "Variable-Rate Low-Delay Analysis-by-Synthesis Speech Coding at 8-16 Kbit/s", Proc. ICASSP 1991, pp. 29-32.

[6] J-H. Chen, N. Jayant, R. V. Cox, "Improving the Performance of the 16 kb/s LD-CELP Speech Coder", Proc. ICASSP, March 1992, pp. I-69.

[7] J-H. Yao, J. J. Shynk, A. Gersho, "Low-Delay Vector Excitation Coding of Speech at 8 Kbit/s", Proceedings of IEEE Globecom'91 Conf., pp. 695-699.

[8] J-H. Chen, M. S. Rauchwerk, "An 8 kb/s Low Delay CELP Speech Coder", Proceedings of IEEE Globecom'91 Conf., pp. 1894-1898.

[9] A. Kataoka, T. Moriya, "A Backward Adaptive 8 kb/s Speech Coder Using Conditional Pitch Prediction", Proc. IEEE Globecom 91 Conf., pp. 1889-1893

[10] A. Husain, V. Cuperman, "Low-Delay Vector Excitation Speech Coding at 8 kb/s", Proc. IEEE Int. Workshop on Intelligient Signal Processing and Communication Systems, March 1992, pp. 149-155.

[11] H. C. Woo, J. D. Gibson, "Low Delay Tree Coding of Speech a 8 kb/s", Proceedings IEEE Globecom 91 Conf., pp. 1884-1888.

[12] R. Pettigrew and V. Cuperman, "Backward Pitch Prediction for Low-Delay Speech Coding", Proc. IEEE Globecom 1989, pp. 34.3.1-34.3.6.

[13] V. Cuperman, R. Pettigrew, "Robust low-complexity backward adaptive pitch predictor for low-delay speech coding", IEE Proceedings-I, vol 138, no 4, August 1991, pp. 338-344.

PART III

SPEECH QUALITY

Methods for assessment of speech quality have been important in the development of high quality low bit rate speech coders. Proper evaluation of voice quality from speech coders is also necessary for setting speech coding standards that can be used in the telephone networks. The standardization activities in digital speech coders have created great interest in subjective evaluation of speech quality. This section includes 3 papers on this important topic. The paper by Dimolitsas provides a comprehensive review of methods recommended for subjective evaluation of digital speech coders. The paper by Martino compares the speech quality of several digital speech coders adopted recently as standards in Europe, North America, and Japan, as part of their digital cellular systems. Finally, the paper by Panzer and Shapley provides a comparative evaluation of a number of methods that are frequently used for assessing subjective quality of speech.

6

SUBJECTIVE ASSESSMENT METHODS FOR THE MEASUREMENT OF DIGITAL SPEECH CODER QUALITY

Spiros Dimolitsas

COMSAT Laboratories, Communications Satellite Corporation,
22300 Comsat Drive, Clarksburg, Maryland 20871-9475, USA.

INTRODUCTION

Standardization activities in digital speech coding over the past few years have resulted in an increasing need to develop and understand the methodologies used to subjectively assess new voice transmission systems before they are introduced into a telephone network. In this chapter a review of subjective methodologies for the assessment of *telephone or good communications quality* digital speech coding systems is provided. Technical aspects concerning the network applications and other characteristics relevant to the type of system under evaluation are briefly considered first, since these factors influence the selection of a suitable assessment methodology. Next, listener opinion tests are described. Finally, articulation and diagnostic tests as well as conversational opinion and field tests are briefly addressed.

Classification of Assessment Methods

The assessment techniques that can be used to quantify the performance of voice communications systems can be classified in several different ways. One approach, attributed to Richards [1], groups methodologies according to their generality into three classes.

The first class comprises those methods that are applicable for assessment of the widest possible range of communication systems. Such methods treat the speech link as an end-to-end network connection in which all normal conditions of use are maintained, or closely reproduced, and all parts perform their normal functions as much as possible. Field and conversation tests [2] are commonly used methods in this class whereby users are interrogated and express opinions on the manner in which their conversation was conducted over the speech connection under evaluation [3].

The second class comprises methods that are less general and time consuming to administer than those of the first class. Listener opinion tests offer a well known example of this approach, whereby speech material is transmitted over the system under investigation and rated by a group of listeners. Articulation and Diagnostic tests, are other examples.

The third class comprises methods that seek to describe relevant fundamental characteristics of the system under evaluation. This class embraces all instrumental (or objective) methods for measuring such system parameters as frequency response and non-linear distortion. Objective measurement methods are discussed briefly in this chapter and in more detail in Reference 4.

Factors Influencing the Selection of an Assessment Method

A number of factors influence the choice of class and type of an assessment method. These include the type of impairments, the environmental and participant conditions, the anticipated quality of the system under evaluation, and the test objectives and cost constraints.

IMPAIRMENTS. The ability to classify the impairments of a communications system can significantly affect the choice of a suitable assessment methodology. Careful consideration needs to be given not only to the nature of the impairments introduced by the systems under evaluation, but also to the specific parameters of interest and the topology of the network into which such systems are to be integrated.

In general, communications impairment factors can be divided into three types, depending on the affected direction of the communication link [1]. The first type includes impairments that cause an increase in listening difficulty when the communications link is unidirectional and no assistance is given to the listener by the talker. The second type comprises impairments that cause difficulty while talking only. Finally, the third type includes impairments that cause difficulty while conversing, or factors associated with the alternation of the talking and listening roles of the participants.

The manner in which these different types of impairments are manifested by communications systems is rarely straight forward, since it typically varies with the particular system and network configuration being considered. For example, digital speech coding systems typically give rise to impairments of the first type, in view of the modeling distortions and quantization noise introduced by the encoding and decoding processes of such systems. Long-haul telephone circuits without echo control can give rise to the second type of impairments. Finally, Digital Circuit Multiplication Equipment (DCME) can introduce impairments of the third type which cause difficulty in conversing.

Mappings such as those just described, which associate general groups of communications systems with one of the three types of impairments are not rigid [5]. However, for the remainder of this chapter only the first type of impairments will be emphasized since digital speech coding systems most often generate degradations that involve difficulty in the listening path only (with the possible exception of coding systems incorporated in DCME, as may arise in future digital cellular systems).

ENVIRONMENTAL & PARTICIPANT FACTORS. In addition to the channel the environment in which the communication is taking place, as well as the characteristics of the participants themselves, must be taken into account when selecting an assessment methodology. For example, an environment containing a high level of ambient noise is expected to introduce impairments of the first type, and a talker might be expected to raise the volume of his or her voice. In this case, it would be important to design an experiment that includes evaluation of the speech codec performance at input levels that exceed the nominal. If, on the other hand, high ambient noise is present in the listener environment only, then it may be more appropriate to design an experiment in which only nominal input levels are considered, but listening takes place in an environment that includes the type of ambient noise expected to be present in the real listener environment. Such conditions may occur, for example, when announcement systems are employed for voice communication to and from mobile vehicles. These examples help to highlight the fact that some *a priori* knowledge with regard to these factors will be important in designing a good experimental subjective evaluation.

EXPECTED CODEC QUALITY. A third factor that can affect the selection of an appropriate evaluation methodology is the anticipated quality of the speech codec (or codec condition) under evaluation. As described later, the suitability and usefulness of the tests defined depend on the range of system quality being evaluated. Probably the most common technique used to determine anticipated speech codec quality relies heavily on the availability of experienced evaluators who can determine the approximate equivalence of a given system with respect to a reference performance scale. Much work has also been done on the objective determination of system quality and speech distortion techniques which can also be used for preliminary determinations of system quality [4].

COST. The final factor that influences the selection of an evaluation methodology is cost. For example, assessment methods belonging to the second class are typically less laborious and less time consuming to administer than those belonging to the first class. Consequently, they also are usually less expensive. (However, the results obtained using methods belonging to the second class are less general than those obtained from the first class). Similarly, assessment methods belonging to the third class are often cheaper to conduct than those belonging to the second class. Therefore, before a particular methodology is selected it is important to estimate the approximate effort involved in conducting tests belonging to each of the three basic assessment classes as well as the desired generality of the results obtained.

SELECTION OF AN ASSESSMENT CLASS

Having classified the nature of telecommunications system impairments into three categories and having also classified assessment techniques into the three classes it is now appropriate to consider the selection of a suitable assessment class once the parameters of interest and the nature of the impairments introduced by the telecommunications environment, network, and speech codec under evaluation have been determined.

An essential feature of the first type of impairments is that only the listening transmission path is involved, and therefore it seems plausible that in many cases appropriate tests can be conducted by using high quality recordings. This avoids the need for live talkers or for a conversation while listening is taking place. Since many of the practical cases involving relatively high quality speech codecs fall into this category, listening only assessment techniques are commonly employed. In certain instances where only a preliminary (as a complement to the second class of methods) or limited assessment is required, the third class of techniques consisting of instrumentation measurements can be employed.

The second type of impairments also affect only one transmission channel, namely the path on which talking is taking place. In this case, methods falling either into the second, or third classes, may be employed. However, because these type of impairments are not generally introduced by digital speech coding systems, they will not be considered further.

The third type of impairments involve difficulty in conversing, and thus require use of one of the more general methods falling under the first class to ensure that the proper conversational structure of speech is present. Either field tests or conversation opinion tests are applicable; the merits of each will be briefly considered later. Before examining specific tests in detail, it is important to consider the application of reference systems, and rating scales.

Reference Systems and Rating Scales

Once an experiment has been conducted, it is desirable to express the information gathered in a simple format, and within a framework that is general enough to serve as a basis for network planning or for comparison with other experimentally obtained databases.

To achieve the desired transportability of results, it is necessary to define some methodology by which the results obtained from a given test can be meaningfully related to the results obtained from another test. To this end, reference systems, standards, and representative connections are used. These comprise speech links composed of stable and specifiable components intended for laboratory or network use, under conditions that enable all essential characteristics to be precisely defined and consistently reproduced at different times and by different laboratories.

These reference systems can be employed to generate conditions that are better than the best and worse than the worst condition to be evaluated, thus "anchoring" the experiment to known performance points. Because the opinion rating scales are not necessarily linear, further anchoring points may be necessary throughout the range of test conditions under study. The most significant and frequently used system for the assessment of digital speech coding has been the Modulated Noise Reference Unit (MNRU) which has been recommended by the International Telephone and Telegraph Consultative Committee (CCITT). Other standardized speech coding techniques such as CCITT Rec. G.711 [6] on 64-kbit/s Pulse Code Modulation (PCM), or CCITT Rec. G.721 [7] on 32-kbit/s Adaptive Differential PCM (ADPCM) can also be used for this purpose.

The MNRU, which is defined in detail in CCITT Rec. P.81 [8], produces random noise with an amplitude that is proportional to the instantaneous speech amplitude (multiplicative noise). This noise is perceptually very similar to quantization noise. The ratio of the speech level to the multiplicative noise level, expressed in dB, is called the Q value. For a given system, an equivalent Q value can be determined by means of subjective tests.

Another useful definition is that of the Mean Opinion Score (MOS). The MOS is a quantifier of subjectively rated transmission performance, which is computed by averaging the individual opinion scores for each circuit condition evaluated by a sample of listeners. These opinion scores represent a listener's assessment of the quality of a speech sample expressed over an appropriately chosen scale. CCITT [9,10], recommends the use of a five-point scale {excellent, good, fair, poor, bad} which is typically numerically mapped to the decimal {5, 4, 3, 2, 1} scale. Listener opinions of performance can also be solicited to assess a number of other transmission performance characteristics, as described in the next section.

In the following table, an example is given of the equivalent Q and MOS subjective performance of three well known coding methods: CCITT Rec. G.728 on 16 kbit/s LD-CELP, CCITT Rec. G.726 on 32 kbit/s ADPCM and CCITT Rec. G.711 on 64 kbit/s PCM. In this table "2 x", and "4 x" denote the asynchronous interconnection of two and four coding methods, respectively.

Type of Coding	*MOS*	*Q Value (dB)*
16 kbit/s LD-CELP	3.93	25.34
2 x 16 kbit/s LD-CELP	3.67	21.98
4 x 16 kbit/s LD-CELP	3.38	19.16
32 kbit/s ADPCM	3.88	24.90
2 x 32 kbit/s ADPCM	3.65	21.75
4 x 32 kbit/s ADPCM	3.34	18.82
2 x 64 kbit/s PCM	3.94	26.04
4 x 64 kbit/s PCM	3.63	21.53

LISTENER OPINION TESTS

Listener opinion tests are conducted using speech material in the form of sentences (typically high quality recordings). The listeners or subjects judge the speech received over the system under evaluation according to a given criterion [12]. Several criteria can be used for obtaining opinion scores including criteria based on loudness preferences, listening effort, degradation with respect to a reference circuit condition, the audibility of transmission or processing noise, and the overall quality of the material listened to.

Listener Opinion Test Objectives

It is important to distinguish between two different objectives that can be achieved when listener opinion techniques are employed. The first is a general one, namely that of obtaining assessment scores that predict the degree of satisfaction likely to be expressed by subjects conversing over symmetrical speech telecommunication links in which the principal paths are identical to the speech codec under evaluation. For such purposes, it is usually undesirable to expose listeners to a succession of samples of speech that are too short or that have been degraded in entirely different ways.

The second aim is associated with the process of developing or improving new speech coding systems. Here, it is often necessary to assess and correctly rate the relative effects of different varieties of one type of degradation. For example, different speech coding quantization schemes may need to be compared. When the expected forms of degradation are similar, sufficient exposure can be obtained during shorter segments of speech, and this objective can often be achieved by employing paired comparison experimental designs.

Types of Listener Opinion Tests

In generating suitable speech material, a set of phonetically balanced sentences uttered by a variety of talkers, both male and female, is normally required. It is common to employ one set of recordings obtained using a microphone appropriate to the various systems under evaluation, and then use the same recordings for several experiments in which the same type of microphone would normally be employed.

In selecting a listening only test, the methods recommended by the CCITT in Rec. P.80 [2] are frequently employed. These listening only methods can be classified into three groups: Absolute Category Rating (ACR), Degradation Category Rating (DCR), or Equality Threshold Rating (ETR). The first is an absolute rating method, while the other two are relative rating methods. Which method is selected is sometimes a matter of individual experimenter preference; however, differences do exist between the methods as discussed next.

ABSOLUTE CATEGORY RATING (ACR). Listener opinion tests administered as ACR assessments are usually conducted by arranging for the listener to hear a succession of groups of typically two to three sentences, each group being reproduced over a different circuit condition and referred to as a sample. After each sample is heard, the listener expresses an opinion in terms of an appropriately chosen scale. Each opinion is based on exposure to the most recently heard sample only. This type of test attempts to force the listener to vote on the most recently heard sample only. Thus, the sample needs to be of sufficient duration to prohibit the listener from making a direct comparison of the current sample with the previous one. At the same time, the sample should not be so long as to result in a lengthy test (e.g., exceeding 1 hour) when many conditions are assessed. Sample lengths of approximately 6 to 10 seconds have been found to be suitable for this purpose.

When the test is completed, the average of all voters' scores for each circuit condition (the MOS) can be calculated. Although it not possible to devise absolute rules governing the use of ACR tests, they have been found to perform well at low Q values ($Q < 20$ dB). However, the usefulness of ACR tests also depends on the type of rating scale used, and each scale has its own range where ACR tests may be appropriate. ACR tests are subject to the "order effect", that is, a fair sample played after a good sample may receive a different opinion score than if the same sample is listened to after a bad sample. This effect can be minimized by choice of a suitable experimental design and by employing an increased number of sample randomizations.

The Q value of a system can be determined subjectively by conducting an experiment where both the system and a number of MNRU references are rated in terms of the MOS. The Q value of the system is subsequently determined by computing the dB value of the MNRU reference that yields the same MOS as the system whose Q value is sought. This process yields the "Q rating" of the unknown system.

DEGRADATION CATEGORY RATING (DCR). An alternative to ACR tests is the DCR technique. The DCR overcomes some of the limitations encountered in ACR tests, particularly the resolution degradation in the good telephony quality range of $Q >$ 20 dB. In the DCR method, a degradation opinion scale is used, and a high quality reference condition always precedes each condition being assessed. In DCR tests, stimuli are presented to listeners as (A–B) pairs, where A is the high quality reference used to anchor each judgment, and B is the same sample after processing. Subjects rate the processed conditions using a degradation scale. Thus, results from DCR tests are usually collected as Degradation MOS, or DMOS. For high quality systems whose performance is similar, DMOS scores tend to be more "spread out" (sensitive) than MOS scores under the same conditions [10].

EQUALITY THRESHOLD RATING (ETR). Equality Threshold Rating (ETR) provides a technique for directly comparing a digital process with the MNRU. This measure leads to a threshold of equality defined as the 50 percent preference level between the MNRU and the digital process (the point where one-half of the listeners prefer one condition and one-half the other). In ETR tests, stimuli are always presented to listeners as (A–B) pairs, where A is the MNRU reference signal and B is the processed sample. Short duration speech segments (2.5 to 5 seconds) are presented to listeners, who are required to make forced judgments on whether A is preferable to B, or vice versa.

PAIRED & RANKED COMPARISONS. Listener opinion tests administered as DCR or ETR assessments involve comparisons against a reference only. Paired comparisons, as well as higher order ranked preference designs, can also be employed; however, these are much less frequently used. Ranked designs share with DCR and ETR test the "relativeness rating" concept, where an evaluated sample is always compared against some other sample (although in this case not a reference sample). In paired comparisons, a test is normally administered by presenting the listener with two samples of two to three sentences each and asking for the listener's binary preference upon completion of a sample pair.

Selecting an Approach

The selection of a suitable experimental approach, particularly the choice between absolute and rank order designs, is very important and is influenced by the type of systems being evaluated, the overall test objective, and the number of total conditions to be assessed. Generally, if the conditions to be assessed are degraded in an entirely different manner, then paired comparison tests (which are relative by nature) are unsuitable for the purpose. Instead, listener opinion ACR MOS tests are preferable, since they are absolute by design. DCR tests are also suitable for this purpose, since only a quantification of the degradation is solicited, and not a forced preference decision. If, on the other hand, the degradation is of the same nature, then paired comparison tests offer an attractive approach.

Similarly, if the goal of the experiment is to evaluate the effect of small differences in performance, then paired comparisons often offer the best approach, since relative assessments tend to be more sensitive to small differences in performance than ACR tests. Such tests are also applicable if the processed conditions are degraded in unusual ways (e.g. contaminated by vehicle background noise). However, caution must be exercised in interpreting the results, since a small difference in subjective performance revealed by relative subjective assessments may not necessarily imply that the degree of user acceptance for the speech codecs being evaluated will be correspondingly different.

Reference Circuit Condition Selection

In listening opinion tests, it is important that at least one group of sentences within each run be heard via a set of "anchor" or reference conditions. Such conditions often consist of a set of MNRU references spanning the Q value range of 0 to 35 dB, in 5 dB increments for example.

Experimental Designs

A variety of experimental designs can be used for listener opinion tests. However, it is common to employ graeco-latin or hyper-graeco-latin squares, in which rows represent listeners, columns represent the order in which conditions are administered, symbols of the first alphabet represent circuit conditions, and symbols of the second alphabet represent talkers and lists of sentences [13]. Although the number of listeners is dependent on the accuracy objectives of the experiment, and will typically vary from experiment to experiment for the same accuracy objective, it is common to employ at least 24 listeners.

Articulation and Diagnostic Techniques

Articulation tests are administered when a quantification of the information capacity of the system under evaluation is required, this being measured by assessing the percentage of correctly recognized speech sounds over the speech path in question. Furthermore, when the quality of the systems under evaluation is rather poor (by public telephone standards), the listener opinion tests may fail to produce reliable results. In this case quality is low and listening effort is high, and a listener to such a system is primarily concerned with just understanding any speech material processed by such systems.

One such test based on the above principle is the Diagnostic Rhyme Test (DRT) [14]. A DRT is a two-choice test in which the listener (usually a trained group) hears a one syllable word from a predefined pair of rhyming words whose initial consonants differ with respect to their phonemic character (such as veal and feel, meat and beat, etc). The listener is then asked to judge which of the two words the speaker has spoken, thereby implicitly indicating whether or not he has apprehended the critical feature.

The DRT has found extensive use for the evaluation of low communications or synthetic quality speech coding systems. Intelligibility scores typically vary in the 70 percent to 92 percent range. For high quality telephone grade systems, intelligibility scores between 92 and 96 percent are normally achievable. However, because such systems usually "clutter" in the mid-90-percent range of performance, it is usually difficult to obtain results with enough statistical significance to distinguish between systems that might perform quite distinctly in an ACR or DCR type of test. For this reason, articulation tests are rarely employed to assess quality in telephone systems.

The Diagnostic Acceptability Measure (DAM) [15] is a listener opinion test that, in addition to the assessment of intelligibility, also provides diagnostic information relative to the pleasantness and overall acceptability of the system under evaluation. The test combines a direct (isometric) and an indirect (parametric) approach to acceptability evaluation.

The DAM test has been used successfully as a tool during the development of new speech codecs. However, as with the articulation tests described above, it is not generally employed to assess user acceptability of telephone quality speech codecs.

CONVERSATION OPINION TESTS

Conversation is the normal mode which telephone connections are used by the general public. Thus, all assessments should ideally be of the conversation type, even though in most cases, this might be impractical due to the labor involved in preparing and administering such tests. Conversation tests are often employed as the next step in new speech transmission technology selection, once an initial selection has been made on the basis of objective or listener opinion tests. To this end, conversational laboratory and field test methods seek to reproduce realistic situations that result in pairs of subjects conversing with each other over the connections to be assessed, and to solicit the participants' reaction either during, or typically after, the completion of their conversation.

Laboratory Conversation Opinion Tests

Laboratory conversation opinion tests are necessary when the impairments present in the link under evaluation cause difficulty in conducting a conversation. These tests are also appropriate if it is suspected that the impairments could affect the outcome of a listener opinion test in a different manner than they might affect a conversation opinion test. Laboratory conversation opinion tests conducted in a controlled environment enable reliable observations to be made, and the effects of different conditions can be more easily assessed than in field tests. However, particular care is necessary to ensure that any effects caused by the artificiality of the situation are minimized [1].

Field Tests

Because the aim of subjective testing is to employ, by as much as possible, end-to-end connections that are as realistic as possible, the use of field tests comprising actual conversations over working telephone connections seems attractive, since the real environment in which these tests take place is undisturbed. Field tests are appropriate when it is suspected that all the conditions and network induced factors that might affect a system's performance might not be known in advance for inclusion in appropriate combinations in a laboratory conversation test.

Field tests are also appropriate when network induced factors are known, but for practical reasons are impossible to reproduce in a laboratory environment. Field tests are useful too when a speech codec's user acceptability is sought, rather than an assessment of a codec's relative performance with respect to other similar systems. Because of cost considerations, field tests are typically used as the final selection mechanism for the adoption of new speech coding technology, and are usually administered after objective, listener opinion, and laboratory conversational tests have been conducted.

Instrumental and Objective Measurement Methods

These methods can be generally classified into two groups. The first group comprises measurement techniques that employ narrowband signals (tones) or are based on the measurement of specific parameters, such as overall noise and loss. The second group consists of methods that directly measure the distortion between primarily digital speech signals. Examples of the first group include measurements of the signal-to-noise ratio, and non-linear distortion with single and multitone signals [16], [17], or calculation of transmission quality and listening effort scores by measuring such parameters as attenuation/frequency distortion, circuit noise, room noise, and sidetone paths [18].

Because many digital speech coders employ an *a priori* knowledge of the speech production and hearing mechanisms, narrowband non-voice signals are generally neither well modeled nor well reproduced. Thus measurements on such signals can be very misleading in predicting speech quality. This indicates the need for different kinds of objective measurement techniques that can be performed by employing speech signals directly. Examples of these techniques, which comprise the second group, include calculation of the signal-to-noise ratio, frequency distortion, cepstral distortion, or maximum-likelihood distortion ratios computed over real speech signals [4].

Conclusions

A brief review of subjective test methodologies applicable to the quality assessment of digital speech coding systems has been presented. For additional in-depth background, the reader is encouraged to consult the P-series CCITT Recommendations as well as D. L. Richards' book [1], which served as the basis for many of the currently recommended test procedures. The reader is cautioned, however, that field of subjective voice assessment is an unusually dynamic one. As a result, continuous reassessment and revision of subjective assessment procedures is being undertaken in order to accommodate new types of degradations arising from the development of new speech coding and transmission technology.

References

[1] D. L. Richards, *Telecommunication by Speech*, New York: John Wiley, 1973.

[2] CCITT, "Methods for Subjective Determination of Transmission Quality, "Revised Rec. P.80, Geneva 1992.

[3] CCITT, "Subjective Performance Assessment of Telephone-Band and Wideband Digital Codecs" Draft Rec. P. 83, Geneva 1992.

[4] S. Dimolitsas, "Objective Speech Distortion Measures and Their Relevance to Speech Quality Assessment," *Proc. IEEE* , Vol. **136**, Pt. I, No. 5, pp. 317–324, October 1989.

[5] Spiros Dimolitsas. "Subjective Quality Quantification of Digital Voice Communications Systems". *IEE Proceedings*, Part **I**: *Communications, Speech and Vision.* Volume **138**, N. 6, pp. 585–595, December 1991.

[6] CCITT, "Pulse Code Modulation (PCM) of Voice Frequencies," Rec. G.711, *Red Book,* Vol. **III.3**, pp. 85–93, Malaga-Torremolinos, 1984.

[7] CCITT, "32-kbit/s Adaptive Differential Pulse Code Modulation (ADPCM)," Rec. G.721, *Red Book*, Vol. **III.3**, pp. 125–159, Malaga-Torremolinos, 1984.

[8] CCITT,. "Modulated Noise Reference Unit (MNRU)," Rec. P.81, *Blue Book*, Vol. **V**, pp. 198–203, Melbourne, 1988.

[9] CCITT, "Subjective Performance Assessments of Digital Processes Using the Modulated Noise Reference Unit (MNRU)," Suppl. 14, *Blue Book*, Vol. **V**, pp. 341–360, Melbourne, 1988.

[10] J. R. Rosenberger, "Quality Assessment Methods for Speech Coding," *Telecommunications Journal*, Vol. **55**, No. 12, 1988, pp. 820–825.

[11] Colin South and Paolino Usai, "Subjective Performance of CCITT's 16 kbit/s LD-CELP Algorithm with Voice Signals," Proceedings, *IEEE Global Communications Conference*, Globecom'92, Orlando, FL, December 1992.

[12] CCITT, "Methods Used for Assessing Telephony Transmission Performance," Suppl. No. 2, *Blue Book*, Vol. **V**, pp. 237–248, Melbourne, 1988.

[13] G. E. P. Box, W. G. Hunter, and J. S. Hunter, *Statistics for Experimenters: An Introduction to Design, Data Analysis and Model Building*, New York: John Wiley, 1978.

[14] W. D. Voiers, "Evaluating Processed Speech Using the Diagnostic Rhyme Test," *Speech Technology*, pp. 30–39, January/February 1983.

[15] W. D. Voiers, "Diagnostic Acceptability Measure for Speech Communications Systems," International Conference on Acoustics, Speech and Signal Processing, Hartford, Connecticut, *Proc. IEEE*, 1977.

[16] *Bell System Technical Reference*, "Transmission Parameters Affecting Voiceband Data. Transmission Measuring Techniques. May 1975," Publication 41009.

[17] S. Dimolitsas, F. L. Corcoran, M. Onufry, and H. G. Suyderhoud, "Evaluation of ADPCM Coders for Digital Circuit Multiplication Equipment," *COMSAT Technical Review*, Vol. **17**, No. 2, pp. 323–345, Fall 1987.

[18] CCITT, "Prediction of Transmission Qualities from Objective Measurements," Suppl. No. 4, *Red Book*, Vol. **V**, pp. 214–236, Malaga-Torremolinos, 1984.

7

SPEECH QUALITY EVALUATION OF THE EUROPEAN, NORTH-AMERICAN AND JAPANESE SPEECH CODING STANDARDS FOR DIGITAL CELLULAR SYSTEMS

Eliana De Martino

DBP Telekom, Research Institute
Darmstadt,Germany

Telebras, R&D Center
13088-061 - Campinas, Brazil P.O. Box 1579
TDMARTINO@CPQD.ANSP.BR

INTRODUCTION

The new generation of cellular telephone systems in Europe, North America and Japan will incorporate digital transmission of speech. Presently there are at least three TDMA digital cellular standard systems: the European (GSM) [1], the North-American (TIA) [2] and the Japanese [3]. A fundamental part of these systems to achieve good speech quality is the speech coding procedure. The three speech coding schemes, RPE-LTP at 13 kbit/s, VSELP at 8 kbit/s and VSELP at 6.7 kbit/s, respectively adopted as GSM, North-American and Japanese standards, have never been compared together in order to give an indication of the difference of performance not only due to the different data rates but also due to the different algorithms themselves. In this paper the subjective speech quality of these three speech coder algorithms is evaluated. The influence of channel errors and the correspondent channel coding for each codec are not considered.

CURRENT STANDARDS ALGORITHMS

The Regular Pulse Excitation-Long Term Prediction (RPE-LTP) speech coding algorithm uses an equi-spaced down-sampling grid to approximate the excitation signal in combination with a long term prediction. For this evaluation the simulation of the RPE-LTP codec is bit exact in line with the GSM Recommendation [1].

The Vector Sum Excited Linear Prediction (VSELP) speech coding algorithm at 8 kbit/s and 6.7 kbit/s is a variation on CELP (Code Excited Linear Prediction) coder [4]. This algorithm uses codebooks with predefined structure to vector quantize the excitation signal reducing the computation in the codebook search process. Two codebooks are used with the VSELP-8 kbit/s and only one with the VSELP-6.7 kbit/s. Because there is no bit exact description of these codecs their simulation is performed based on the functional description found in the literature [2, 3]. Table 1 shows the data rate distribution of the three standards including channel coding.

Details	EUROPEAN	N. AMERICAN	JAPANESE
Speech coding algorithm	RPE-LTP	VSELP	VSELP
Speech coding data rate	13.0 kbit/s	8.0 kbit/s	6.7 kbit/s
Channel coding data rate	9.8 kbit/s	5.0 kbit/s	4.5 kbit/s
Gross data rate	22.8 kbit/s	13.0 kbit/s	11.2 kbit/s

Table 1 - Data rate distribution

The evaluation made here takes into account only the most important criterion to analyse the performance of a speech coder scheme: the speech quality. In particular, the comparison is concentrated on the basic speech quality of the codecs not considering the robustness of each one to channel errors. Table 2 shows the data rate distribution of the speech coding parameters for the three algorithms.

Parameters	RPE-LTP	VSELP 8 kbit/s	VSELP 6.7 kbit/s
Short-term prediction	1.8	2.15	2.1
Long-term prediction	1.8	1.4	1.4
Excitation	9.4	4.4	3.2
Net data rate	13.0	7.95	6.7

Table 2 - Speech coding parameters (kbit/s)

TEST CONDITIONS

A formal subjective testing was carried out to assess the performance of the speech codecs. The test was conducted to evaluate the sensitivity of the codecs to input levels (12, 22, 32 dB below overload point of the codec) and to different talkers (2 male and 2 female). The recording environmental noise was lower than 30 dBA and the active speech level measure was made using a speech voltmeter conforming to CCITT Recommendation P.56 [5]. The speech material was in german language and consisted of elements of two sentences with a duration of about 2 seconds each one separated by 2 seconds of silence. The speech samples are weighted according to the sending side of an Intermediate Reference System as specified in CCITT Recommendation P.48 [6]. The listening level was -10 dBPa in an environment noise of lower than 45 dBA. The experiment was broken into four segments each one having a random order of presentation. A group of 18 non-expert listeners took part in the test. The evaluation was based on the mean opinion score (MOS) values, using as reference system the MNRU as proposed by CCITT [7]. The experiment has used for the MNRU a range of correlated noise ratio (Q) from 5 to 40 dB in steps of 5 dB. For each condition of the codecs and for each MNRU condition were used 4 different elements - one for each speaker - giving a total of 68 different elements.

TEST RESULTS

The test results are showed in Figure 1: a) is the MNRU curve, b) is the overall result (average male and female voices), c) is the result for male voice and d) is the result for female voice. Table 3 shows the confidence interval obtained in the test for the overall result.

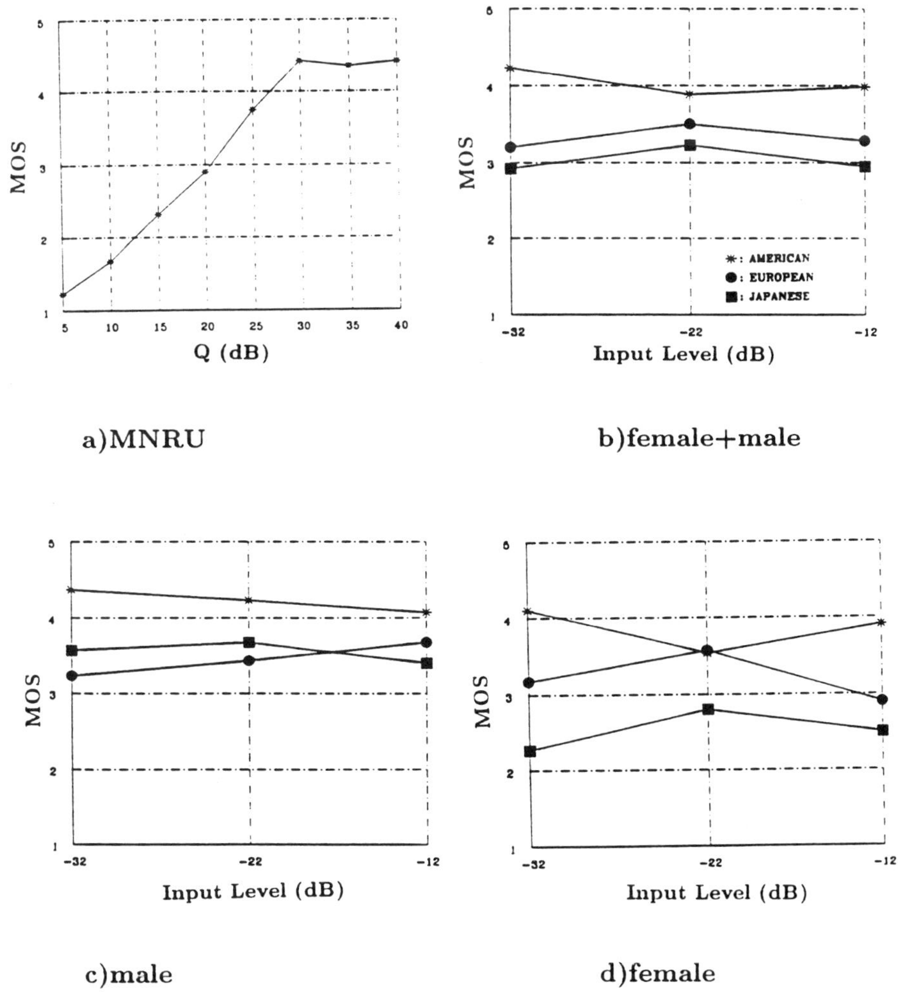

Figure 1 - Test results

From Figure 1 and Table 3, it is evident that, despite the lower data rate, the basic speech quality of the VSELP-8 kbit/s was better or equivalent in all conditions to the RPE-LTP codec. The VSELP-6.7 kbit/s showed a basic speech quality statistically equivalent to the RPE-LTP, although a lower performance was obtained for female speakers.

Input Level(dB)	-12	-22	-32
European	±0.16	±0.18	±0.16
American	±0.17	±0.21	±0.17
Japanese	±0.25	±0.23	±0.25

Table 3 - Confidence interval for overall results

CONCLUSIONS

The results of this comparison give an indication of the different speech quality that can be found on the future digital cellular system taking into account only the differences of performance of the speech coding procedure. These results are restricted to the limitations of the test itself which did not include the effect of tandem and channel errors. A comparison of the different DMR systems in a realistic operating environment including channel errors would require a common channel-model taking the different gross bit rates into account.

REFERENCES

[1] GSM, "GSM Full Rate Speech Transcoding", Rec. 06.10, 1988.

[2] Electronics Industries Association, "Cellular System", Report IS-54, 1989.

[3] Motorola, "Vector Sum Excited Linear Prediction (VSELP) 11200 bit per second voice coding algorithm including error control for Japan Digital Cellular", Draft text for specification, 1990.

[4] M.R.Schroeder, B.S. Atal, "Code-Excited Linear Prediction (CELP): High quality speech at very low bit rates", Proc. Int. Conf. on Acoustics, Speech and Signal Proc., pp 937-940, 1985.

[5] CCITT, "Objective measurement of active speech level", Rec. P56, Blue Book, Vol V, 1989.

[6] CCITT, "Specification for the Intermediate Reference System", Rec. P48, Blue Book, Vol V, 1989.

[7] CCITT, "Modulated Noise Reference Unit (MNRU)", Rec. P81, Blue Book, Vol V, 1989.

8

A COMPARISON OF SUBJECTIVE METHODS FOR EVALUATING SPEECH QUALITY

Ira L. Panzer, Alan D. Sharpley, and William D. Voiers

Dynastat, Inc.
2704 Rio Grande, Suite 4
Austin, Texas 78705

INTRODUCTION

With the advances realized in voice coding algorithms over the past two decades it has become increasingly evident that speech intelligibility, alone, is not a sufficient criterion of system performance. As a result, a number of methods have been developed to measure the quality or acceptability of speech. Several methods have been used fairly extensively. These include, in particular, the Diagnostic Acceptability Measure (DAM), which reports a Composite Acceptability Estimate (CAE), the Absolute Category Rating (ACR) method, which reports a Mean Opinion Score (MOS), and the Degradation Category Rating (DCR) method, which reports a Degradation Mean Opinion Score (DMOS). Comparison of these methods, based solely on data in the literature, is difficult, if not impossible. Given the many recent developments in speech coding technology for network and wireless applications, there is a clear need for a rigorous comparative evaluation of the major methods of acceptability evaluation. The purposes of this investigation were (1) to examine the interrelations among scores yielded by three methods of evaluating speech acceptability and (2) to compare the resolving powers of these methods with several types of coincidental and systematic speech degradation commonly encountered in modern digital voice communications.

METHODS FOR MEASURING SPEECH QUALITY

Diagnostic Acceptability Measure (DAM)

The DAM, developed by Voiers [1], has several unique features. First, it combines a direct (isometric) and an indirect (parametric) approach to acceptability evaluation. In addition to rating acceptability of a speech sample, directly, listeners also have the opportunity to indicate, independently, the extent to which various perceived qualities are present in the sample, without regard to how these qualities may affect acceptability. For example, two listeners may disagree on their overall acceptability ratings of a speech sample with background noise, while agreeing on the amount of noise present in the sample.

A second feature of the DAM is the requirement that listeners make separate ratings of the speech signal itself, the background, and the total effect. Listeners make a total of 21 ratings during the course of a speech sample. Ten ratings are concerned with perceptual qualities of the signal, eight ratings are concerned with the perceptual qualities of the background, and three ratings are concerned with perceived intelligibility, pleasantness, and overall acceptability. A summary of these rating scales is shown in Fig. 1. An example of a typical response screen is shown in Fig. 2. The 21 ratings are combined to compute a CAE for reporting acceptability. (How these ratings are combined and the additional scores produced by the DAM are beyond the scope of this paper.)

PARAMETRIC SIGNAL RATINGS
- S1 Fluttering/Pulsating
- S2 Dull/Muffled
- S3 Rasping/Rough
- S4 Small/Distant
- S5 Babbling/Slobbering
- S6 Thin/Tinny
- S7 Scratchy/Dry
- S8 Nasal/Whining
- S9 Interrupted/Chopped

ISOMETRIC SIGNAL RATING
- S_i Unnatural/Distorted

METAMETRIC OVERALL RATINGS
- I_i Intelligibility
- P_i Pleasantness

PARAMETRIC BACKGROUND RATINGS
- B1 Hissing/Fizzing
- B2 Chirping/Clicking
- B3 Rushing/Roaring
- B4 Bubbling/Percolating
- B5 Crackling/Staticy
- B6 Rumbling/Rolling
- B7 Humming/Buzzing

ISOMETRIC BACKGROUND RATING
- B_i Conspicuous/Intrusive

ISOMETRIC OVERALL RATING
- A_i Overall Acceptability

Fig. 1. DAM II rating scale descriptors.

System number xx	***SIGNAL***	
For this condition the	OVERWHELMING	(9)
SPECIFIC characteristic	EXTREMELY CONSPICUOUS	(8)
of the SPEECH SIGNAL	VERY CONSPICUOUS	(7)
described as	QUITE CONSPICUOUS	(6)
	RATHER CONSPICUOUS	(5)
	MODERATELY NOTICEABLE	(4)
	SOMEWHAT NOTICEABLE	(3)
	SLIGHTLY NOTICEABLE	(2)
RASPING	BARELY DETECTABLE	(1)
ROUGH is:	NOT DETECTABLE	(0)

Fig. 2. Replica of listener monitor showing a sample DAM rating scale.

A third unique feature of the DAM is the concept of the normative listener. Listeners bring different "yardsticks" (i.e., different subjective origins and scales) to the task of making acceptability ratings. To compensate for these differences, listeners are independently calibrated on a standard set of speech materials and their rating data are compared to those of a hypothetical normative listener, and appropriately transformed to approximate those of the normative listener.

Absolute Category Rating Method (ACR)

The ACR, described by Goodman [2,3], is both easy to implement and convenient. It requires listeners to make a single isometric rating of each speech sample. A typical response screen for the ACR is shown in Fig. 3. For purposes of computing an MOS, the letter responses are converted to a five point numeric scale.

System number xx		
Which category best	EXCELLENT	- E
describes the system	GOOD	- G
you just heard for	FAIR	- F
purposes of everyday	POOR	- P
telephone communication?	BAD	- B

Fig. 3. Replica of listener monitor showing a sample ACR rating scale.

Degradation Category Rating Method (DCR)

The DCR is a method used by Combescure, *et al* [4] in an attempt to provide improved resolution among good telephone quality coders. It requires the listener to make a single, comparative rating of annoyance as shown in Fig. 4. The rating results in a reported DMOS.

System number XX	
Which category best	5 - Degradation is inaudible
describes the second	4 - Degradation is audible but not annoying
system you just heard	3 - Degradation is slightly annoying
compared to the first	2 - Degradation is annoying
system?	1 - Degradation is very annoying

Fig. 4. Replica of listener monitor showing a sample DCR rating scale.

MATERIALS

The speech materials consisted of 30 systems selected from a large customer database of DAMs previously evaluated by Dynastat. These systems were specifically selected to provide a range of speech quality across several standard speech coding algorithms and two systematic degradations. For this study 15 systems were nominally characterized as narrowband systems (2400 and 4800 bps coders in the quiet, noise, and with bit-errors) and the remaining 15 as wideband systems (CVSD, ADPCM, N-bit PCM, and MNRU [5]). The processed speech material for each system consisted of 12 six-syllable sentences (approximately 48 seconds) for each of three male and three female speakers.

With the DAM, listeners rate all systems for one speaker and then cycle to the next speaker. For each speaker the materials are presented in a different order to soften the possible effects of context. For present purposes, the 12 processed sentences for each system were dubbed onto DAT from the original recordings provided by customers, resulting in six 30 system DAMs.

The ACR materials were generated by digitally dubbing from the DAM tapes two sentences for each system. Thus, the presentation order and level remained constant across methods. Each set of two sentences was followed by a four second listener response interval.

The DCR materials were generated by digitally dubbing from the ACR tapes, however, each of the two sentences was proceeded by a reference condition. In the $DCR_{(+30)}$ materials, the reference was +30dB MNRU, one of the systems to be evaluated. In the $DCR_{(High)}$ materials, the reference was high fidelity speech. Also, in $DCR_{(High)}$ only one sentence was dubbed from the ACR tape. In both sets of DCR materials the sentences were followed by a four second listener response interval.

In Experiment 1, the 30 systems described above were evaluated using the three methods by 20 members of Dynastat's listening crew used in normal test operations. Listeners were seated in a sound isolation room in front of individual microcomputers. Materials were presented diotically over TDH-39 elements at 87dB SPL as measured on Dynastat's audio distribution system. In Experiment 2, the two sets of DCR materials were presented to another, independent set of 19 listeners.

RESULTS

Under the circumstances of this investigation, i.e., a common set of systems evaluated by all methods involved, the F-ratio [6] for "systems" provides a valid indication of relative resolving power. Thus, the F-ratios in Table 1 indicate that DAM and DCR-1 provided essentially equal resolution across all 30 systems

Table 1. F-Ratios from Analyses of Variance for Experiments 1 and 2.

	Experiment 1			Experiment 2	
	DAM	ACR	$DCR\text{-}1_{(+30)}$	$DCR\text{-}2_{(+30)}$	$DCR\text{-}3_{(High)}$
All Systems	95.7	71.6	96.7	110.7	72.4
Narrowband Systems	61.1	40.3	30.1	40.3	33.9
Wideband Systems	87.7	62.7	102.0	120.2	78.4
MNRU Systems	107.0	123.7	175.4	213.9	131.5
PCM Systems	95.1	66.3	61.8	105.7	69.5
Other Systems	111.7	63.2	80.6	98.9	56.6

involved. However, an examination of various subsets of systems, indicates that the methods varied significantly in their ability to resolve system differences. For example, in narrowband systems, the DAM shows the greatest resolution among the three methods, while, in wideband systems, the DCR shows the greatest resolution. However, it should be noted that DCR-1's F-ratio varies dramatically across the subsets of wideband systems. Also, as shown by the results for Experiment 2, the DCR's ability to resolve among systems is a function of the reference system selected. DCR with the *high* reference has consistently lower F-ratios than DCR with the *+30dB* reference.

To facilitate graphic comparisons among the methods, all scores were converted to a common scale, such that the average variance of system-mean scores was equal to one and the grand mean of the system scores was zero. Thus, the slopes of all graphs relating test scores to level of <u>systematic</u> degradation are proportional to the resolving powers of the methods involved. Shown first, however, (Figs. 5-7) are the interrelations among scores for the three methods. In Fig. 5 the relation between CAE and MOS appears to be virtually linear. In Figs. 6 and 7, however, a curvilinear relation is evident when DMOS-1 is plotted

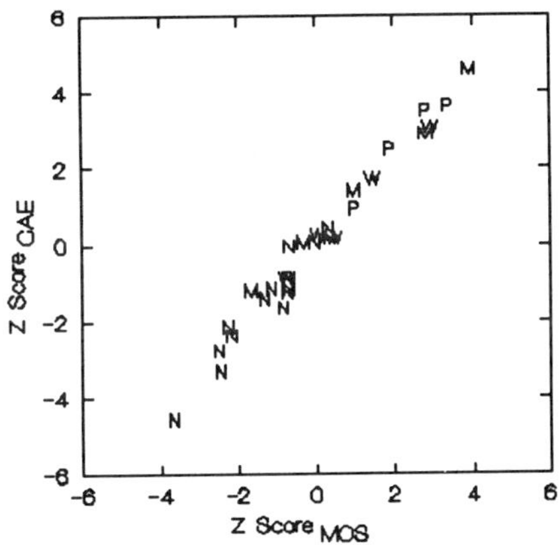

Fig. 5. CAE vs. MOS for all systems in experiment 1.

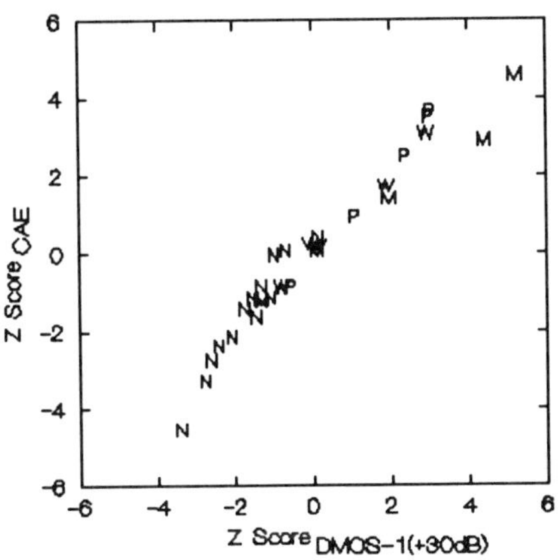

Fig. 6. CAE vs. DMOS-1$_{(+30)}$ for all systems in Experiment 1.

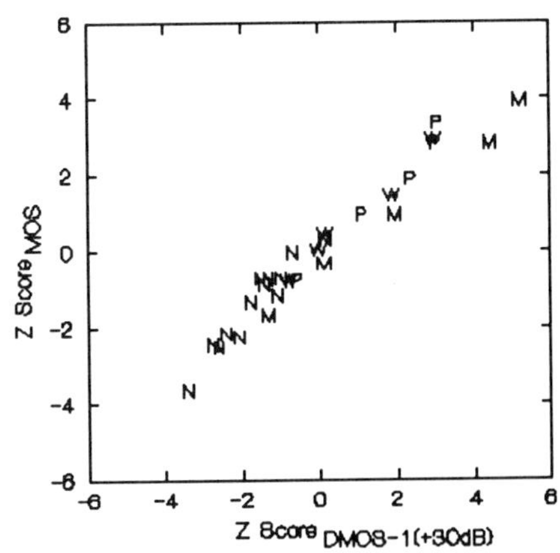

Fig. 7. MOS vs. DMOS-1$_{(+30)}$ for all systems in Experiment 1.

against CAE and MOS, a result of ratings on the MNRU conditions (points coded with an M).

In Fig. 8, where the MNRU conditions are plotted for each of the three methods, the DMOS-1 results for +24dB and +30dB diverge from the linear trends shown by DAM and MOS.

Figure 9 shows that, by changing the DMOS reference condition to high fidelity speech (Experiment 2), a linear trend emerges for MNRU conditions. Figure 10 shows the n-bit PCM conditions plotted for each method. One would expect the DMOS to diverge, as in the case of MNRU conditions, given that MNRU conditions are used to simulate the effects of quantization. However, Fig. 11 shows little difference between DMOS-2 and DMOS-3.

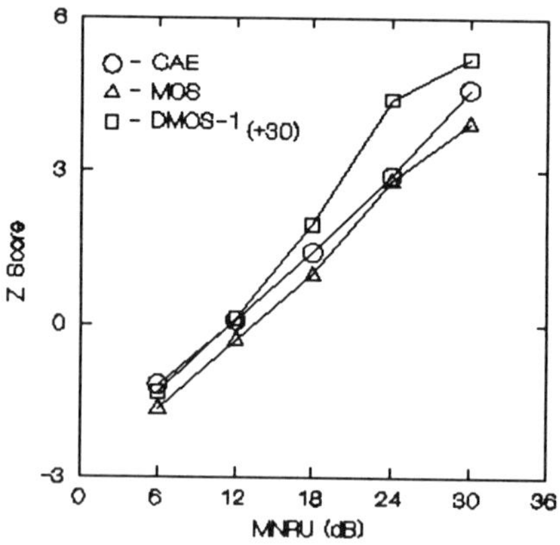

Fig. 8. Z scores vs. MNRU for experiment 1.

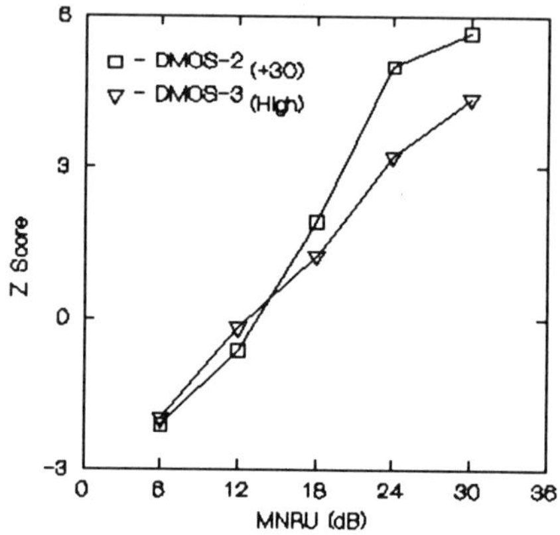

Fig. 9. Z scores vs. MNRU for experiment 2.

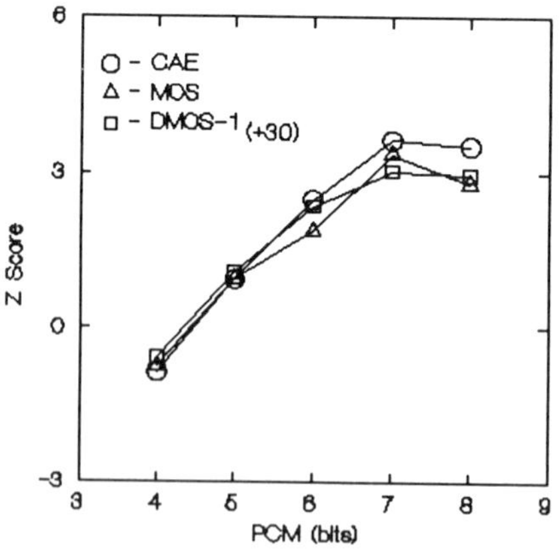

Fig. 10. Z scores vs. N-bit PCM for experiment 1.

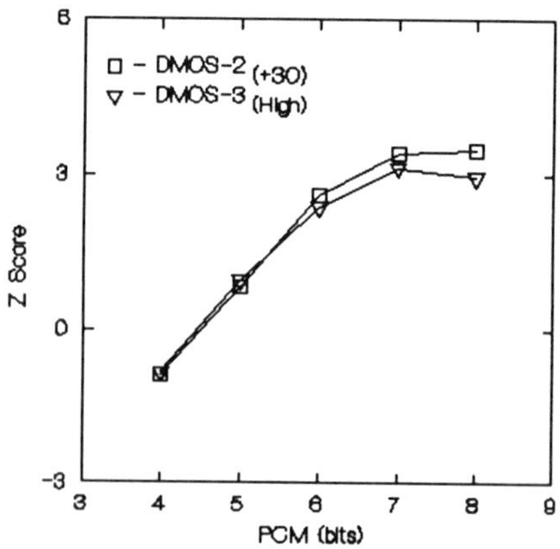

Fig. 11. Z scores vs. N-bit PCM for experiment 2.

The above results confirm that the ability of the DCR method to resolve among systems is a function both of the reference condition and the systems involved. The DCR-$1_{(+30)}$ provided the best resolution among MNRU conditions, but the worst resolution for the PCM and narrowband conditions. This raises the issue of how the DCR scale should be used when evaluating systems that differ in many dimensions from the reference. Depending upon the reference selected, listeners may be responding to the audibility, and thus their annoyance, of some signal quality (i.e., distortion) or some background quality (i.e., noise) independent of the effect these degradations have on overall speech quality.

SUMMARY

Data from these experiments indicate that although the three methods investigated are highly correlated, they do provide varying degrees of resolution, depending on the class of systems involved. Given the importance of speech quality among the many criteria used to compare speech coding algorithms, additional research on the appropriate use of the testing methods is clearly warranted.

REFERENCES

[1] Voiers, W.D., "Diagnostic Acceptability Measure for Speech Communication Systems," *Proceedings of the IEEE International Conference on Acoustics, Speech, and Signal Processing*, Hartford, CT, May 1977.
[2] IEEE Subcommittee on Subjective Measures. "IEEE Recommended Practice for Speech Quality Measurements," *IEEE Trans. Audio and Electroacoustics*, 17, pp. 227-246.
[3] Goodman, D.J. and R. Nash, "Subjective Quality of the Same Speech Transmission Conditions in Seven Different Countries," *IEEE Transactions on Communications*, Vol. COM-30, No. 4, April 1982.
[4] Combescure, P., A. LeGuyader, and A. Gilloire, "Quality Evaluation of 32Kbit/s Coded Speech by Means of Degradation Category Ratings," *Proceedings of the IEEE International Conference on Acoustics, Speech, and Signal Processing*, Paris, France, May, 1982.
[5] Law, H.B. and R.A. Seymore, "A Reference Distortion System using Modulated Noise," *Proc. Inst. Elec. Eng.*, Vol 109B, pp.484-485, 1966.
[6] Guilford, J.P., *Fundamental Statistics in Psychology and Education*. New York: McGraw Hill, 1965.

PART IV

SPEECH CODING FOR WIRELESS COMMUNICATIONS

Low-rate speech coding is a major ingredient of the emerging digital cellular networks. While the first systems based on the TIA IS-54 digital cellular standard in North America, the Japanese digital cellular (JDC) system, and the GSM standard in Europe are deployed, systems which will allow doubling of the capacity - half-rate digital cellular - are already in design and three new standards for Europe, North America, and Japan are nearly finalized. This section starts with a paper by Su and Mermelstein which presents a possible candidate for the North-American half-rate digital cellular competition. A candidate to the half-rate GSM competition is presented by Dervaux, Gruet, and Delprat.

The IS-54 and the GSM digital cellular standards are based on Time Division Multiple Access (TDMA). A competing digital cellular system based on Code Division Multiple Access (CDMA) under consideration by the TIA as an alternate standard will use variable rate speech coding. The chapter contributed by Paksoy and Gersho represents a useful introduction to variable rate speech coding. Jacobs and Gardner present in their chapter the main contender for variable rate speech coding in a CDMA environment.

Most speech codecs for half-rate digital cellular work at a rate of about 4 kb/s (another 2.4 kb/s are allocated for channel coding and 1.6 kb/s for control information, giving an aggregate rate of 8 kb/s). LeBlanc and Cuperman present new results in multi-stage vector quantization of the speech spectral parameters for 4 kb/s speech coding. Finally, Kleijn and Granzow present waveform interpolation, a new technique for speech coding at low rates which shows promise at 4 kb/s.

9

DELAYED DECISION CODING OF PITCH AND INNOVATION SIGNALS IN CODE-EXCITED LINEAR PREDICTION CODING OF SPEECH

Huan-yu Su and Paul Mermelstein

Bell-Northern Research Ltd.
16, Place du Commerce
Verdun, Québec, Canada H3E 1H6

INTRODUCTION

Most linear prediction speech codecs (coder-decoders) employ a fixed frame duration for linear prediction computation, which is a compromise between the rate of spectrum variation of the speech signal and the transmission requirements of the LPC information. The LPC residual signal is coded by considering subframes significantly shorter than the LPC frame. Such subframe-based excitation computations are motivated by considerations of computational complexity. Global search through the space of excitations for the entire frame does not increase coding delay, but requires computational resources beyond those available on one or two chips today.

Mano and Moriya [1] have proposed use of an (M,L) tree-coding procedure for LPC residual coding in CELP. We provide results for a delay-preserving formulation of the solution. Excitation hypotheses are generated subframe by subframe. An unique coding decision is forced at the end of each frame.

In code-excited linear prediction speech coding [2], encoding of the pitch residual is achieved by vector quantization (VQ), and the pitch predictor is realized by either an adaptive long-term filter [2,3], or by an adaptive vector quantizer [4,5]. The determination of the pitch predictor parameters and the optimal innovation signal is the objective of a perceptually weighted minimization process, based on an analysis-by-synthesis closed-loop search procedure. Figure 1 illustrates the typical CELP encoder schema.

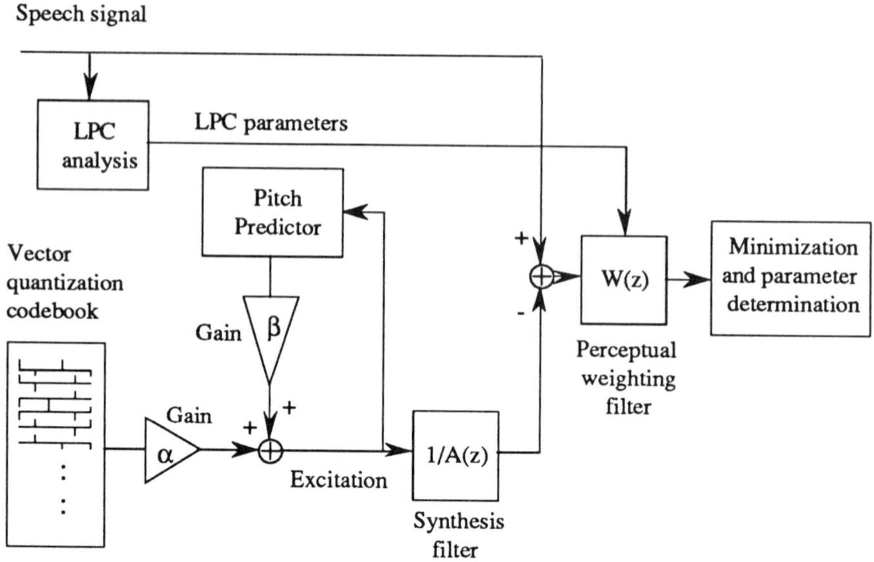

Figure 1. CELP based speech encoder

Most proposed speech codecs operating at medium to low bit-rates (4.8 - 16 Kbits/sec), perform the LPC analysis once per frame (10 - 40 ms). Pitch analysis and vector quantization, on the other hand, are performed once per subframe (2 - 8 ms). In other words, the minimization of the global perceptually weighted quantization error is replaced by a series of lower dimensional minimizations over disjoint temporal intervals:

i) For each subframe, the pitch analyzer determines the optimal pitch parameters (lag and gain β), taking into account the ringing of the synthesis filter, 1/A(z), and the ringing of the perceptual weighting filter, W(z), from the signals generated in the previous subframe;

ii) Then the VQ analyzer determines one optimal VQ parameter set (index of the optimal innovation codevector and its associated gain α). The ringing and the contribution of the pitch predictor (defined by the pitch parameters) are both considered as fixed additive components in the minimization procedure.

This subframe-based minimization procedure is the most efficient in terms of reducing the computational requirements. It is, however, not optimal in terms of approaching a global minimum. Three principal factors account for this: first, subframes are processed independently; second, pitch contribution and the VQ contribution are evaluated separately; and third, preforming pitch analysis before VQ analysis implies that pitch prediction is the more important contributor to the error minimization. In next section, we will study in detail these three points.

The suboptimal error minimization yields adequate performance at medium bit-rates (8-16 Kbits/sec), since the quantization error is small enough to provide good speech quality [5]. At low bit-rates (4-8 Kbits/sec), the residual error is larger and the deficiencies of the approach become more significant.

DELAYED DECISION CODING

To improve CELP codec performance at low bit-rates, improvements in the minimization procedures are needed. The delayed decision coding represents a good alternative [1]. Before discussing the advantages of this technique, let us review the pitch prediction operations.

When the pitch predictor is realized as a one-level adaptive VQ process, past excitation signals are used to construct an adaptive codebook for the current subframe, within the bounds of the range permitted in pitch lag. As we know that past excitation signals are determined by the selections of past pitch parameters and VQ parameters, the contents of the adaptive codebook (or the pitch prediction) are affected by the previous parameter selections. This is the main problem of the subframe-based minimization procedure. In other words, because of the nature of pitch prediction (feed-back), independent parameter selections from subframe to subframe do not necessarily lead to a global minimum in weighted error over the frame.

One may expect that the performance of the pitch prediction for the current subframe would be improved if the current prediction error were taken into account in past analyses. One other observation is that within perceptually important regions (sustained voicing, onset, voicing transitions, etc.), the contribution of pitch prediction to the excitation energy is usually more than 50%, and can reach more than 90% for sustained voiced segments. Hence, the perceptual quality is likely to improve with better pitch prediction.

For a given subframe, the weighted error energy is computed as:

$$\varepsilon = \| \mathbf{D} - \beta \mathbf{E}_{Lag}\mathbf{H} - \alpha \mathbf{C}_i \mathbf{H} \|^2$$

where $\mathbf{D}$ is the perceptually weighted input speech signal from which the ringing of the filter W(z)/A(z) has been subtracted (Figure 2), $\mathbf{E}_{Lag}$ is the pitch contribution with its gain β, $\mathbf{C}_i$ is the VQ contribution with its gain α, and $\mathbf{H}$ is the impulse responses matrix of W(z)/A(z). The parameter selection consists in finding Lag, β, i and α so that ε is minimized. Joint minimization requires that all possible combinations of these parameters be taken into account, a computationally

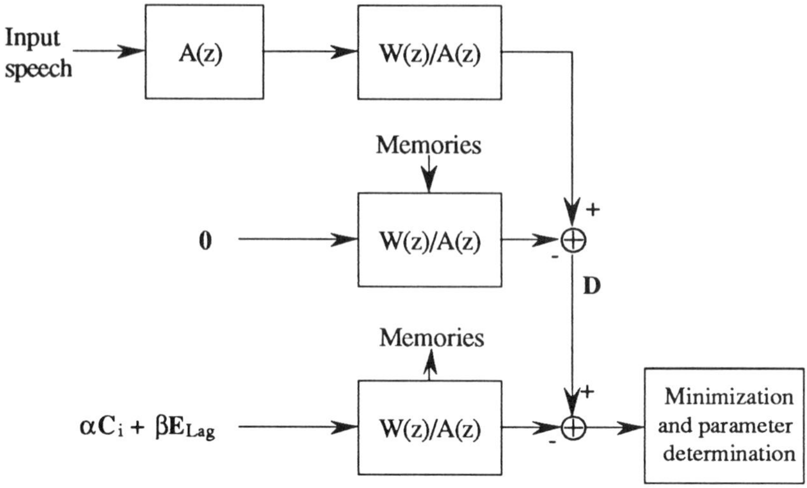

Figure 2. Modified minimization in a CELP coder

prohibitive task. Again, suboptimal methods have to be used. Since the pitch contribution is more significant in perceptually important regions, most techniques perform the pitch analysis first, minimizing the following error expression by selecting the pitch vector components $\mathbf{E}_{Lag}$ and β,

$$\varepsilon' = \| \mathbf{D} - \beta \mathbf{E}_{Lag} \mathbf{H} \|^2$$

Once $\mathbf{E}_{Lag}$ and β have been determined, the VQ parameter set is determined by minimizing the total error:

$$\varepsilon = \| (\mathbf{D} - \beta \mathbf{E}_{Lag} \mathbf{H}) - \alpha \mathbf{C}_i \mathbf{H} \|^2$$

This two-stage minimization procedure gives good results when the pitch contribution dominates the VQ contribution (for example, exceeding 80% of the excitation energy). But in transition and onset regions, the VQ contribution can be more important than the pitch contribution. Thus, an intermediate method between the full joint minimization and the totally separated minimization is worthy of investigation.

Delayed decision coding tends to overcome the disadvantages of subframe-based minimization mentioned before by using an (M,L) tree coding technique: multiple candidates are preserved at each analysis level so that a list of coding hypotheses is carried through the coding frame. The final decision regarding excitation in one subframe is not made until subsequent subframes have been considered. Coding performance is greatly improved but the computational requirements increase exponentially with the number of hypotheses considered.

To avoid additional speech coding delay and to reduce the processing requirement, we introduce a fixed delay level coding technique:

i) for the first subframe, the N_p best pitch candidates are saved by the pitch analyzer; for each of these pitch candidates, a VQ analysis is performed and the N_c best VQ candidates are kept. Thus $N_p \times N_c$ possible excitation signals are computed, but only N_{max} best candidates are saved;

ii) for the second subframe, there are N_{max} possible past excitation signals, so N_{max} pitch analyses are performed and $N_{max} \times N_p$ best pitch candidates are saved. $N_{max} \times N_p \times N_c$ possible excitations are computed, but again only N_{max} of them are saved;

iii) the same procedure is repeated for each subframe, except for the last;

iv) for the last subframe, the same procedure is repeated, but only one optimal excitation signal is saved for use in the next speech frame. The coding of the current speech frame is then completed without additional coding delay.

PERFORMANCE EVALUATION

Implementation complexity is primarily dependent on N_{max}, but also increases with N_p and N_c. Objective performance can be studied as N_p, N_c and N_{max} are varied. Below we give results for reasonable values of the search parameters.

Figure 3 shows the perceptually weighted SNR performance of the delayed decision coding technique. An improvement of the order of 1.1 dB is achieved in SNR, and 0.9 dB in SNRSEG when $N_p=N_c=N_{max}=10$. The speech codec used for this evaluation is operating at 4.7 Kbits/sec with a randomly populated codebook, and the perceptual weighting filter $W(z)=A(z)/A(z/\gamma)$ with a factor $\gamma=0.8$ [2]. All parameters are coded. 10 sentences from 10 speakers (5 male and 5 female, 26 seconds speech) comprised data base for the evaluation.

Similar improvement in terms of SNR can be observed with a codec operating at 4 Kbits/sec, however, the improvement in perceptual quality is more evident than in the case of 4.7 Kbits/sec.

We observe that delaying only the pitch decision results in little improvement with N_p larger than 4. The number of alternative physically reasonable pitch hypotheses can be readily restricted to a small value. Preserving more pitch candidates at each pitch analysis step is more likely to result in a lower minimum for each subframe after the VQ analysis. But a pitch candidate, which by itself does not yield good prediction gain, is unlikely to form an optimal combination with a VQ candidate. Thus N_p will typically be in the range of 2 to 4, while the N_{max} and N_c do not manifest saturation until 10.

In figure 4, signal waveforms are shown in speech onset region. Delayed decision coding achieves a significant improvement in reconstructing the natural waveform as compared to the classic subframe-based coding.

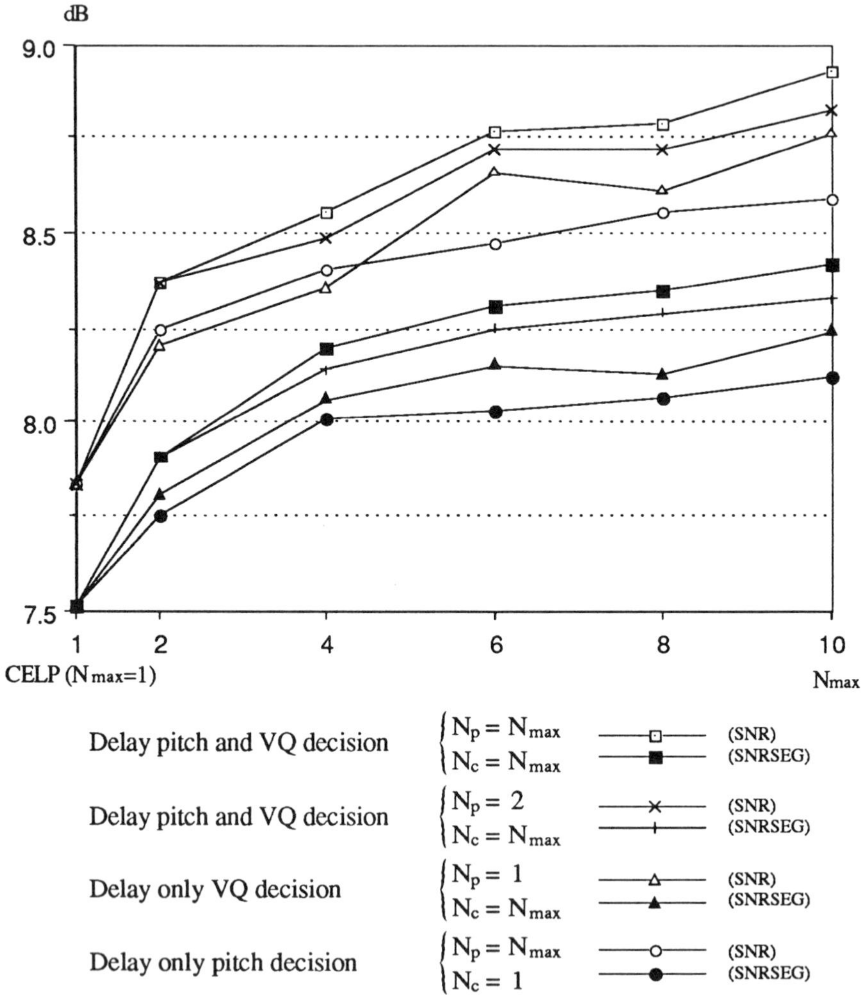

Figure 3. Perceptually weighted SNR performance of delayed decision coding. Frame duration = 40 ms, subframe duration = 8 ms

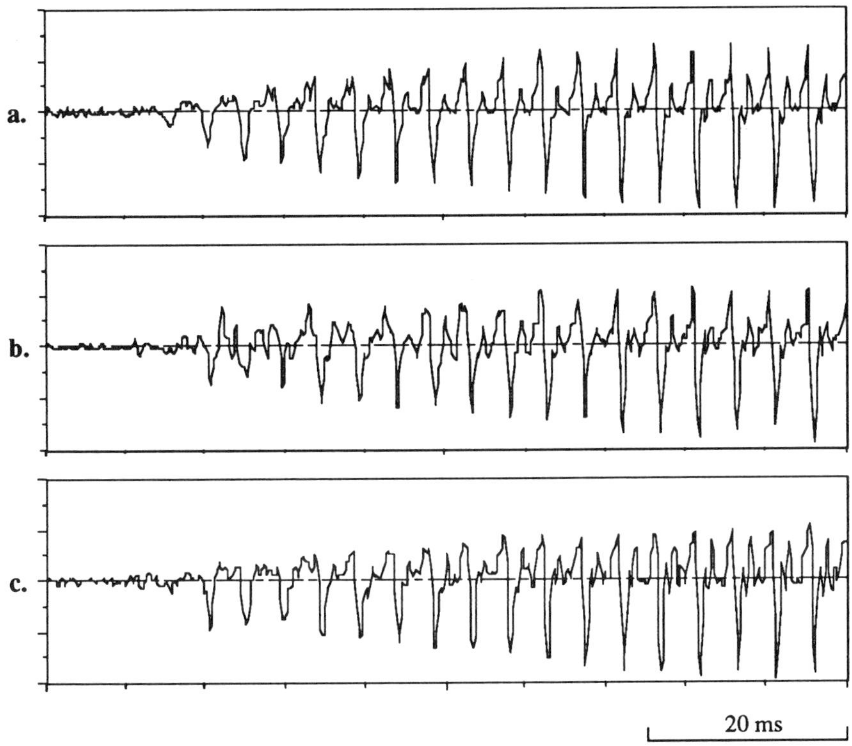

Figure 4. Time domain signal waveforms: a. input speech signal, b. output speech signal using subframe-based minimization, and c. output speech signal using delayed decision coding.

CONCLUSIONS

Code-excited linear prediction employing both short-term and long-term (pitch) predictors involves execution of a complex optimization procedure. To achieve speech codecs of reasonable complexity, generally requiring no more than one DSP chip, we need to simplify the optimization process. In this chapter we presented an alternative approach to the optimization procedure which improves speech quality of low bit-rate codecs at the cost of a moderate increase in complexity. The technique shows significant potential for enabling high quality speech coding at bit-rates of 4 Kbits/sec and below.

Improvements are particularly apparent in regions where the speech excitation signal is changing rapidly such as regions of voicing onset. While coding decisions regarding the individual subframes are delayed beyond the subframe boundaries, they are not delayed beyond frame boundaries. This avoids increasing the overall coding delay, an important requirement for codecs deployed on telephone networks.

REFERENCES

[1] Kazunori Mano and Takehiro Moriya, *4.8 KBIT/s Delayed Decision CELP Coder Using Tree Coding,* Proceedings of ICASSP, pp.21-24, 1990

[2] M.R. Schroeder and B.S. Atal, *Code-excited linear prediction (CELP): High quality speech at very low bit rates,* Proceedings of ICASSP, pp.937-940, 1985

[3] J.P. Adoul, et al, *Fast CELP coding based on algebraic codes,* Proceedings of ICASSP, pp.1957-1960, 1987

[4] D. Lin, *Speech Coding Using Efficient Pseudo-Stochastic Block Codes,* Proceedings of ICASSP, pp.1354-1357, 1987

[5] I. Gerson and M. Jasiuk, *Vector sum excited linear prediction (VSELP) speech coding at 8 Kb/s,* Proceedings of ICASSP, pp.461-464, 1990

10

VARIABLE RATE SPEECH CODING FOR CELLULAR NETWORKS†

Allen Gersho and Erdal Paksoy

Center for Information Processing Research
Department of Electrical and Computer Engineering
University of California, Santa Barbara, CA 93106

INTRODUCTION

A central objective in the design of a cellular network for mobile or personal communication is to maximize capacity while maintaining an acceptable level of voice quality under varying traffic and channel conditions. Conventional FDMA and TDMA techniques, dedicate a channel or time slot to one unidirectional speech signal regardless of the fact that a speaker is silent roughly 65% of the time in a two-way conversation. Furthermore, when speech is present, the short-term rate-distortion trade-off varies quite widely with the changing phonetic character. Thus, the number of bits needed to code a speech frame for a given perceived quality varies widely with time. The speech quality of coders operating at a fixed bit rate is largely determined by the *worst-case* speech segments, i.e., those that are the most difficult to code at that rate. Variable rate coding can achieve a given level of quality at an average bit-rate R_a that is substantially less than the bit rate R_f that would be required by an equivalent quality fixed rate coder. Efficient multiple-access systems, such as CDMA, directly translate this rate reduction into a corresponding increase in network capacity.

Variable rate coders can be divided into two main categories: (a) **source-controlled** variable rate coders, where the coding algorithm responds to the time-varying local character of the speech signal to determine the data rate, and (b) **network-controlled** variable rate coders, where the coder responds to an external control signal to switch the data rate to one of a predetermined set of alternative rates. The external control signal is assumed to be remotely generated, typically in response to traffic levels in the network or in response to requests for signaling information.

In source-controlled coding, the coder in some fashion dynamically allocates bits in response to the local (short-term) character of the speech source. Such coders are intended to maintain a desired level of quality for each short segment of speech with the fewest bits needed. Coders that exploit voice activity patterns to code active speech segments at a fixed rate and silent segments at a reduced rate (or zero rate)

† This work was supported in part by the National Science Foundation, Fujitsu Laboratories, Ltd, the UC Micro Program, Rockwell International Corporation, Hughes Aircraft Company, and Eastman Kodak Company.

are important members of the class of source-controlled coders. Many other source-controlled techniques can be used to code active speech segments with variable rate. One important approach to source-controlled variable rate coding is based on phonetic classification of speech segments where a different rate (and coding procedure) is used for different classes. Such coders can readily include voice activity detection as an integral part of the phonetic classification stage.

Network-controlled variable rate coders can be viewed as **multi-mode** variable rate coders, where a different mode of encoding or perhaps an entirely distinct coding algorithm is performed for each bit rate option. A special case of particular interest is an **embedded** coder, where a single coding algorithm generates a fixed-rate data stream from which one of several reduced rate data signals can be extracted by a simple bit-dropping procedure. The corresponding decoder fills in the missing bits with zeros and then decodes the resulting (modified) full-rate data signal with a fixed decoder algorithm. Thus, each lower rate data signal is *embedded* in the bit stream of the next higher rate data signal.

In this chapter, we outline some of the issues of variable rate coding relevant to cellular networks. In particular, we consider both voice activity patterns and phonetic segmentation of speech. Finally, we briefly examine three multiple-access systems which benefit from variable rate coding.

EXPLOITING VOICE ACTIVITY PATTERNS

In a classic study of voice activity patterns, Brady [1] observed that one side of two way telephone conversations consists of intermittent *talk spurts* separated by pauses or silence. The process of identifying when talk spurts occur is called *voice activity detection* (VAD), *speech activity detection*, or simply *speech detection*. Based on a simple speech detector, Brady found that the average speaker is talking about 44% of the time. Subsequent studies based on more sophisticated detectors of voice activity have found a lower percentage of active talking time. In particular, Yatsuzuka reported the value of 36% based on a sensitive and sophisticated voice activity detector [2].

The quality of the VAD algorithm is a very important consideration in the design of systems that enhance capacity by exploiting voice activity. The increase in capacity is determined by the *voice activity factor* (VAF) which is the fraction (or percent) of the time the detector identifies the presence of active speech. Reliably measuring the VAF of a detector requires averaging over one side of the conversation in many calls with many different speakers. If silence is detected as speech, the capacity is reduced; on the other hand, when speech is detected as silence, degradations in the recovered speech quality are introduced.

During speech pauses, the acoustic signal is not really "silence". Background noise, at some level, is always present. The task of a VAD algorithm is complicated because certain speech sounds have a very low energy level and are random in character and thereby often confused with background noise. Difficult phonemes (phonetic units of speech) for a detector include weak fricative sounds such as /f/ in *fat* and /h/ in *hat*. There are often special situations where a detector could incorrectly or prematurely declare the start of a silent interval. To avoid detecting extremely brief pauses (which may cause excessive overhead in some multiple-access systems) and to reduce the risk of audible clipping due to premature declaration of a silent interval when the background noise level is very high, some *hangover time* is desirable. During the hangover time, the VAD delays its decision and continues to observe the waveform before it declares that a transition has occurred from active speech to silence. While this decreases the VAF, it helps to reduce degradations resulting from the algorithm's inability to make reliable decisions in very noisy environments.

VAD decisions are usually based on multiple features extracted from the audio signal including time varying energy, zero-crossing counts, sign bit sequences, and features generated from within the speech coding algorithm. For the mobile environment, the design of a VAD is complicated by the high level of acoustic noise reaching the microphone. To avoid degrading the speech quality, the VAD algorithm can be designed fairly conservatively so that a lot of the background noise will be classified as active speech rather than silence. This, however, can increase the VAF from 35% or 40% to as high as 60%, reducing the potential capacity gain.

To preserve the naturalness of the recovered speech signal for the listener, it is generally desirable to reproduce in some fashion the background noise. The original noise can either be coded at a very low bit rate or statistically similar noise, called *comfort noise*, can be regenerated at the receiver. Studies of VAD for cellular applications are given in [3], and [4].

VARIABLE RATE CODING OF ACTIVE SPEECH

Variable rate coding of *active* speech segments is a natural way to achieve further reductions in average bit-rate. In this section, we briefly review various approaches to source-controlled variable rate algorithms for coding of speech segments that have been declared as active by a VAD algorithm. For brevity, we omit consideration of variable length coding (e.g., Huffman coding) and focus on ways that are more explicitly based on the speech characteristics.

High quality fixed rate coding of speech at medium bit rates can be achieved using analysis-by-synthesis predictive coding algorithms such as *code-excited linear prediction* (CELP). When the rate is pushed below 4 kbit/s, the performance of CELP algorithms tends to degrade rapidly. Often, the degradation in reproduced speech is caused primarily by a small fraction of frames, containing phonetic units of speech that are more critical than average and hence require a higher than average bit rate to encode adequately. This motivates allocating varying number of bits to different speech segments.

The large dynamic range of the short-term power of speech can lead to variable rate coding strategies where more bits are allocated to the high energy frames than to low energy frames. One such example is QCELP, a variable rate coder based on CELP that is part of a proposed wideband digital cellular standard [5]. In QCELP, each 20 ms frame may have one of four different basic rates, 8, 4, 2, and 1 kbit/s. The coder selects one of the four rates for each frame by comparing the energy level of the frame with a set of three adaptive threshold levels. This is a form of VAD, where generally the lowest rate is assigned to speech pauses and the highest rate to active speech. In this way background noise during pauses is coded and reproduced at the receiver; this eliminates the need for comfort noise and reduces the degradation that would otherwise be caused when active voice frames have energy below the lowest threshold. The two intermediate rates are used relatively infrequently and typically correspond to marginal cases where the presence or absence of voice is less easily discriminated.

A variable rate CELP coder with six coding states was proposed by Vaseghi [6] where variable rate coding is done by switching between different CELP type coders that vary only in their bit allocation and overall rate. The selected state for each frame is based on the prior state and the character of the current input.

A more sophisticated approach to variable rate coding is to classify speech segments into phonetically distinct categories so that a meaningful decision about the needed bit rate and the coding mechanism can be made for each class [7]. A study of the various types of distortion introduced in different phonetic segments of speech by various speech coders revealed that inadequate adaptation to changing phonetic content is indeed a major limitation of the prevailing family of CELP algorithms. Hence, it seems that a coder that would closely monitor the waveform to be coded and employ coding strategies tailored to individual phonetic groups, could be much more efficient in terms of the rate/quality trade-off.

In Phonetically Segmented VXC (PS-VXC) [8] each coding frame is analyzed to determine a set of features which are then used to phonetically classify the given frame. Tests conducted on large speech files show that after silence elimination approximately 65% of the speech frames correspond to voiced speech, around 30% are unvoiced and 5% can be classified as *onsets*, transitions from unvoiced to voiced speech.

The number of bits required for each phonetic category varies widely. For instance, unvoiced segments do not need a long-term predictor and coarser quantization of both the short-term predictor and stochastic excitation will not significantly affect the quality of the synthesized speech. To illustrate this, we tested the performance of PS-VXC by reducing the excitation and LPC codebook sizes for unvoiced segments. Informal listening tests show that the perceived quality of coded speech remains the same when the number of bits used to encode the unvoiced segments is reduced drastically. These results were confirmed by MOS estimates* obtained using the Bark Spectral Distortion (BSD) measure proposed in [9]. Table 1 shows the MOS estimates obtained using the BSD. Note that the predicted MOS remains essentially constant as the rate is lowered. These rates are for active speech without silence and are not intended to represent the performance of a fully designed variable rate coder.

* The MOS estimates obtained using the BSD are slightly lower than expected but are more accurate in a relative sense, with the estimated MOS score of 4.15 for PCM taken as a reference.

Coding Rate				MOS Estimate
unvoiced (b/s)			average (b/s)	
excitation	LPC	total		
1467	800	2267	3060	2.82
1467	367	1833	2930	2.82
800	800	1600	2860	2.80
800	367	1167	2625	2.83

Table 2 : Performance of PS-VXC at various average coding rates

The bit rate can also be reduced for sustained vowel sounds whose formants and pitch exhibit a slow temporal variation, resulting in a large amount of interframe correlation of the spectrum and the pitch. This correlation can be exploited using differential encoding and appropriate interpolation of the short-term and long-term predictor parameters. On the other hand, allocating more bits to relatively infrequent but perceptually important onsets can offer a significant improvement in quality with only a very modest increase in the average bit rate. Another variable rate coder using phonetic segmentation was also presented in [10] where it was reported that an average bit rate between 5 and 6 kbit/s yields only a small degradation over a 9.6 kbit/s fixed rate coder.

It is important to stress that a coder based on phonetic segmentation operates on perceptually meaningful principles rather than simpler criteria such as signal energy or quantization SNR. Since the ultimate objective is to produce coded speech with good perceptual quality, a phonetic segmentation approach shows considerable promise for low rate coding. It appears that a trend in favor of phonetic segmentation is already emerging: most of the candidate coders in the TIA half-rate assessment have incorporated a multilevel voicing decision allowing up to four voicing categories for a frame. This can be regarded as a form of phonetic segmentation. However, in these coders, the bit rate is necessarily fixed as required for the TDMA application. Nevertheless coders based on phonetic segmentation are particularly well-suited for variable-rate coding.

NETWORK CONTROL OF BIT RATE

Generally, it is much simpler to design a variable rate coder whose rate is externally controlled rather than source controlled. The simplest way is to design a family of fixed-rate coders each with a different rate and simply switch to the appropriate coder to produce the currently needed rate. For frequent rate switching, (e.g., once per frame rather than once per minute), a smooth transition may require preserving the context or the state of the old coder for use in initializing the new coder. In such cases, each coder is likely to use the same algorithm but with modified bit allocations. For infrequent switching it is possible to have entirely different coders.

Embedded coding offers a more elegant approach to external rate control. Since the coder itself generates a fixed rate stream, rate switching is simply achieved by suitable bit-dropping. Embedded coding first became of practical interest for ADPCM where a simple and effective method is available for achieving graceful degradation of quality as the rate is dropped. Recently, a method for achieving embedded coding in CELP type coders was introduced by Iacovo and Sereno [11]. Source-controlled variable rate coders can readily be modified to include network control of the rate by simply forcing the coder to switch to one of the rates normally selected by source control. The QCELP coder includes a network-controlled rate feature, where an external control signal can request the 4 kbit/s rate for a frame in order to send signaling information.

MULTIPLE-ACCESS SYSTEMS WITH VARIABLE RATE CODING

Variable rate speech coding can be efficiently incorporated into suitable multiple-access cellular systems. We first consider systems that exploit speech activity patterns for increased capacity and then consider how variable rate coding of active speech can be utilized.

The GSM standard, based on TDMA with slow frequency hopping, includes an option for *discontinuous transmission* (DTX) where interference (and power consumption) are reduced when individual mobiles stop transmitting during pause segments. By reducing interference, capacity can be increased. The VAD method for this standard is described in [4].

It is possible for certain cellular systems to perform statistical multiplexing of talk spurts to increase capacity by adapting an approach used for some time in cable and satellite multichannel systems. A much larger number of simultaneous calls can be accommodated over a fixed number of circuits by time sharing each one way voice circuit in a multiplexed digital carrier system with talk spurts from different calls. Such schemes are commonly known as digital speech interpolation (DSI) systems. Packet voice techniques for wired networks also have potential for wireless applications. All such schemes exploit speech activity patterns.

Three different multiple-access schemes, E-TDMA, PRMA, and CDMA, which benefit from variable rate coding with VAD are outlined next. The first two can be viewed as more or less directly applying the DSI concept, while the third achieves the capacity gain from variable rate coding in a somewhat different manner.

A recent proposal for applying DSI to TDMA systems called *enhanced* TDMA or E-TDMA was introduced by Hughes Network Systems. In E-TDMA, mobiles are not assigned a slot for the duration of a call but are dynamically assigned slots in a group of full duplex RF channels. Each channel contains six half-rate time slots so that for a 12 channel group there are a total of 72 slots of which 63 are available for talk spurts and 9 are reserved for control overhead data needed to track the location of talk spurt assignments for each voice signal. Thus, for example, when a particular mobile's talk spurt ends, the slot in which it was assigned is vacated and can be assigned to a speaker from any other mobile unit in the group with a newly starting talk spurt. Unlike DSI systems for cable or satellite transmission, the bit-rate is constant during active speech, enabling the future half-rate TIA speech coding algorithm to be adapted to E-TDMA. On the other hand, a fixed rate for active speech will presumably not allow the same graceful degradation during heavy instantaneous talk spurt activity in the DSI group as is attainable when network-controlled rate variation is used. The design will have to be adequately conservative to avoid excessive degradation due to front-end clipping. Also to avoid excessive capacity for overhead data, the VAD may require a long hangover time and will not exploit short pauses between talk spurts.

Another application of VAD and variable rate coding of active speech is in packetized voice transmission. Fast packet switching systems for T1 transmission and asynchronous transfer mode (ATM) for broadband fiber systems have motivated interest in applying packetized speech transmission to wireless networks. A packetized speech scheme called *packet reservation multiple access* (PRMA) which uses VAD was proposed for cellular TDMA systems by Goodman [12]. PRMA, originally conceived for local area wireless networks, dynamically assigns a sequence of fixed-length packets corresponding to one talk spurt to a TDMA slot. At the start of a talk spurt, the terminal contends for an available slot. Once assigned, the slot is reserved for the duration of a talk spurt. When the talk spurt has ended, the slot is released. By including suitable control information in the packet headers, distributed network control is achievable.

CDMA offers a natural and easy way to benefit from variable rate coding in cellular networks. Reducing the bit rate of a speaker correspondingly reduces the interference to other users. Since each user transmits a wideband signal covering the entire assigned spectral band for the service, there is no family of RF channels and no assignment of talk spurts to different channels. Thus, the rather complex overhead associated with DSI systems or the overhead due to packet headers in packet transmission are eliminated. In Qualcomm's CDMA proposal to the TIA [5], each frame may have one of four rates and the receiver automatically identifies the rate without requiring side information. A CDMA system can also be enhanced by the addition of traffic controlled rate variation where the network directs all transmitters to use a lower rate during periods of heavy traffic.

SUMMARY

Variable rate coding of speech for cellular networks is an inevitable direction for future generations of digital cellular and microcellular networks. Initially most attention will be devoted to the use of voice activity as the means to achieve variable rate. Eventually, we expect that network-controlled as well as source-controlled variable rate speech coding will be an important additional feature for very efficient high capacity multiple-access systems.

References

[1] P. T. Brady, "A technique for investigating on-off patterns of speech," *Bell Syst. Tech. J.*, vol. 44, pp. 1-22, 1965.

[2] Y. Yatsuzuka, ''Highly sensitive speech detector and high-speed voiceband data discriminator in DSI-ADPCM systems,'' *IEEE Trans Commun*, vol. 30, pp. 739-750, April 1982.

[3] S. Hatamian, ''Enhanced speech activity detection for mobile telephony,'' *Proc. IEEE Vehicular Technology Society Conf.*, vol. 1, pp. 159-162, Denver, May 1992.

[4] D.K. Freeman, G. Cosier, C.B. Southcott, and I. Boyd, ''The voice activity detector for the Pan-European digital celluar mobile telephone servcie,'' *Proc. Int'l. Conf. Acoust. Speech, Sig. Proc.*, vol. 1, pp. 369-372, Glasgow, May 1989.

[5] Qualcomm, Inc., ''Proposed EIA/TIA Interim Standard - Wideband Spread Spectrum Digital Cellular System Dual-Mode Mobile Station - Base Station Compatibility Standard,'' *Submitted to the TIA TR45.5 Subcommittee,*, April 21, 1992.

[6] S. V. Vaseghi, ''Finite state CELP for variable rate speech coding,'' *IEE Proc.-I*, vol. 138, pp. 603-610, December 1991.

[7] Shihua Wang and Allen Gersho, ''Improved phonetically-segmented vector excitation coding at 3.4 kbit/s,'' *Proc. of the IEEE Int. Conf. on Acoust., Speech, and Sig. Proc.*, pp. 49-52, San Fransisco, March 1992.

[8] Shihua Wang and Allen Gersho, ''Phonetically-based vector excitation coding of speech at 3.6 kbit/s,'' *Proc. of the IEEE Int. Conf. on Acoust., Speech, and Sig. Proc.*, pp. I-349-352, Glasgow, May 1989.

[9] S. Wang, A. Sekey, and A. Gersho, ''Auditory distortion measure for speech coding,'' *Proc. of the IEEE Int. Conf. on Acoust., Speech, and Sig. Proc.*, pp. 493-496, Toronto, Canada, May 1991.

[10] R. Di Francesco, C. Lamblin, A. Leguyader, and D. Massaloux, ''Variable rate speech coding with online segmentation and fast algebraic codes,'' *Proc. of the IEEE Int. Conf. on Acoust., Speech, and Sig. Proc.*, pp. 233-236, Albuquerque, New Mexico, April 1990.

[11] R. D. De Iacovo and D. Sereno, ''Embedded CELP coding for variable bit-rate between 6.4 and 9.6 kbit/s,'' *Proc. of the IEEE Int. Conf. on Acoust., Speech, and Sig. Proc.*, pp. 681-683, Toronto, Canada, May 1991.

[12] D. J. Goodman, ''Cellular Packet Communications,'' *IEEE Trans Commun*, vol. 38, pp. 1272-1280, August 1990.

11

QCELP: A VARIABLE RATE SPEECH CODER FOR CDMA DIGITAL CELLULAR

William Gardner, Paul Jacobs and Chong Lee

QUALCOMM Inc.
10555 Sorrento Valley Road
San Diego, CA 92121, USA

INTRODUCTION

Digital cellular telephone systems require efficient encoding of speech to achieve capacity improvements required of the next generation of cellular systems. The use of a variable rate speech coder can reduce the average data rate required to transmit conversational speech by a factor of two or more, while providing many other advantages. This reduction in average data rate leads to a factor of two increase in the capacity of a Code Division Multiple Access, or CDMA, based digital cellular telephone system by decreasing the mutual interference among users. This chapter describes "QCELP," a variable rate speech coder which has been selected as the speech coding algorithm for the TIA North American digital cellular standard based on CDMA technology.

NEXT GENERATION CELLULAR SYSTEM REQUIREMENTS

Due to the rapid growth of the cellular industry in North America, the Cellular Telecommunications Industry Association has specified that the next generation cellular technology provide a 10-fold increase in capacity, increased coverage and improved quality over the current analog cellular system [2]. To achieve this, the industry has adopted digital technology.

The initial North American digital cellular standard, IS-54 [3], is based on Time Division Multiple Access, or TDMA, technology. This system uses a speech coder, VSELP, which encodes at a fixed data rate of 8 kbit/s. With this coder, IS-54 can achieve no better than a 3-fold increase over the current analog capacity. In order to better meet the industry's requirements, the CTIA has requested a new digital cellular standard based on CDMA spread spectrum technology. The CDMA system proposed in [1] has many advantages over TDMA systems, including: virtually no undetected channel errors, seamless support of many user services (e.g. multiple speech coders), natural implementation of variable data rates without the use of complicated time/frequency slot allocation algorithms and their associated signaling overhead, efficient use of many forms of diversity, a high frequency reuse factor requiring no frequency planning, and soft handoff (i.e. make before break) capability. In large scale field

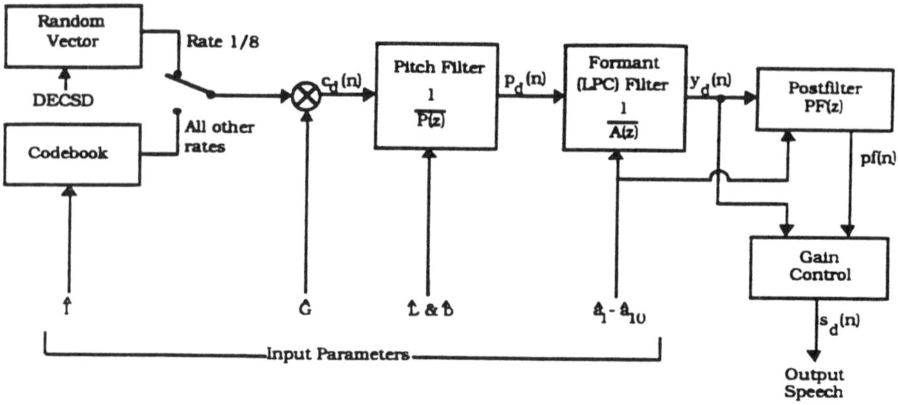

Figure 1: The QCELP Decoder

tests, the proposed system has demonstrated capacity improvements averaging 15 times analog capacity under loaded conditions while providing excellent voice quality [4].

THE QCELP SPEECH CODER

QCELP, the speech coding algorithm proposed in [1], dynamically selects one of four data rates every 20 ms, depending on the speech activity. The four data rates used are 8 kbit/s ("full rate"), 4 kbit/s ("half rate"), 2 kbit/s ("quarter rate"), or approximately 1 kbit/s ("eighth rate"). Typically, active speech is coded at the 8 kbit/s rate, while silence and background noise are coded at the lower rates. MOS testing has shown that QCELP provides speech quality equivalent to that of 8 kbit/s VSELP, while maintaining an average data rate under 4 kbit/s in a typical conversation. This section briefly describes the overall structure of the QCELP algorithm. Figure 1 shows a block diagram of the decoder. A complete specification of the algorithm can be found in reference [1].

The QCELP speech coder is based on the Code Excited Linear Prediction, or CELP, analysis-by-synthesis structure first introduced in [5]. The basic structure of QCELP is scalable, which minimizes complexity by allowing an integrated implementation of all 4 rates. At higher rates, the LPC parameters are more finely quantized and the pitch and codebook parameters are updated more frequently. The exact bit allocation for each data rate is shown in Figure 2. For example, the LPC parameters are updated once per frame, using 40 bits at full rate, 20 bits at half rate, and 10 bits for the two lowest rates. The pitch parameters are updated using 10 bits at a rate of four, two, one and zero times per frame for full, half, quarter and eighth rate, respectively. Similarly, the codebook parameters are updated using 10 bits at a rate of eight, four and two times per frame for full, half and quarter rate, respectively. 6 bits are used for the one codebook parameter update in eighth rate frames, as described below.

A 10th order LPC filter is used. Its coefficients are encoded using LSP fre-

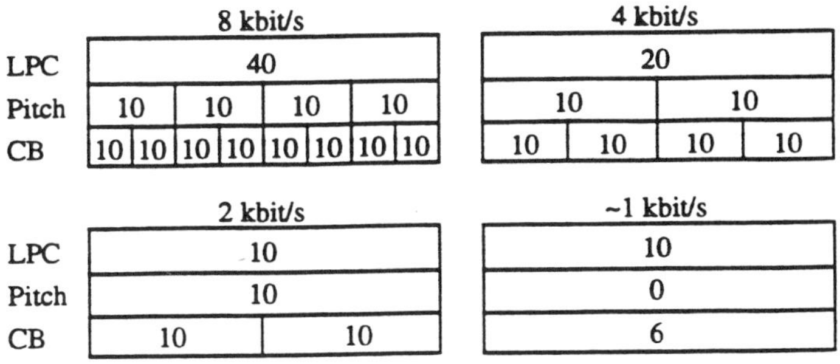

Figure 2: Bit Allocations for Each Data Rate

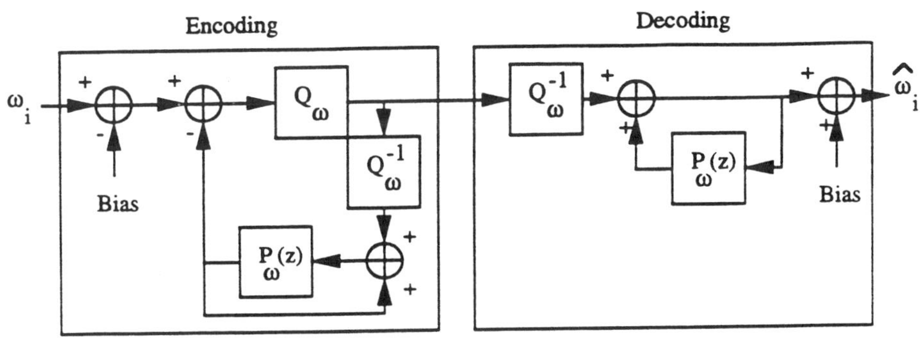

Figure 3: LSP Quantization

quencies [6] due to the good quantization, interpolation, and stability properties of LSPs. Each LSP frequency is coded using a differential quantizer, shown in Figure 3. First, the bias of each frequency, which is the value the frequency would take on if the input were truly "white noise," is subtracted. The difference between the resulting value and a predicted value based on the previous frame is then scalar quantized. Each rate uses a different scalar quantizer, with a different dynamic range and quantization step size, for each LSP frequency. The predictor $P_\omega(z)$ is $0.9z^{-1}$. Differential encoding of the frequencies exploits the interframe correlation of the LSPs and allows accurate reproduction of arbitrary tones.

The pitch filter has the form

$$\frac{1}{P(z)} = \frac{1}{1 - bz^{-L}}$$

where b is the "pitch gain" and L is the "pitch lag." The pitch lag is quantized from 17 to 143 samples using 7 bits for each pitch update. Due to the large number of codebook updates per frame the use of fractional pitch lags was found to provide little improvement, and thus fractional pitch is not used. The pitch gain is scalar quantized from 0 to 2 using 3 bits once per pitch parameter update.[1] L and b are chosen by standard analysis-by-synthesis error minimization procedures described in [1]. Recursive convolution techniques are used to reduce the complexity. In order to perform the recursive convolution, the output of the pitch filter is needed both from the previous subframes and the current subframe. Since the output for the current subframe has not yet been determined, the "formant residual" (the input speech filtered by $A(z)$) is used as an estimate. To determine the optimal b and L, a global search is performed over all allowable quantized values of L and b, rather than over all L with the quantization of b performed after the search as is traditionally done. This global search ensures that the truly optimal b and L are found, and gives slightly improved performance.

The index of the codebook I and the gain factor G are determined once for each codebook update. A gaussian, center-clipped, recursive codebook of length 128 is used. As in the pitch search procedure, I and G are chosen using analysis-by-synthesis procedures. Due to the high update rate of codebook parameters, there is significant correlation in the codebook gain which allows the gains to be encoded differentially. For each codebook update, the sign of G is transmitted using 1 bit, and the log of the magnitude of G is differentially encoded using only 2 bits, using a quantizer similar to that used for the LSP frequencies. As in the pitch search procedure, the search is performed over all allowable quantization levels of I and G.

The coder structure is modified slightly during eighth rate frames to code background noise more efficiently. Because the pitch filter provides no improvement in background noise, the pitch gain is set to zero for all eighth rate frames. In addition, the codebook index and the sign of the codebook gain are not transmitted, and the codebook itself is replaced by a white noise generator. The seed for the generator is a function of the eighth rate packet of data, which is available at both the encoder and the decoder. This ensures that both the encoder and decoder produce the same random noise sequence, keeping them synchronized.

The decoder uses an adaptive postfilter of the form

$$PF(z) = B(z)\frac{1 - \sum_{i=1}^{10} \alpha^i a_i z^{-i}}{1 - \sum_{i=1}^{10} \beta^i a_i z^{-i}}$$

where $\alpha = .5$, $\beta = .8$, and the a_is are the LPC filter coefficients. $B(z)$ is an adaptive brightness/dimness filter, which compensates for the spectral tilt created by the postfilter. The amount of spectral tilt is estimated using a function of the average of the 10 LSP frequencies. Unity gain through the postfilter is maintained with an AGC.

---

[1] If the pitch gain is 0, the lag is irrelevant. Because of this, a zero pitch gain is encoded by setting the pitch lag to 16. The decoder checks for this special case, and sets b to zero if it receives a lag of 16. Thus, there are 9 possible levels for the pitch gain, and 127 possible pitch lags.

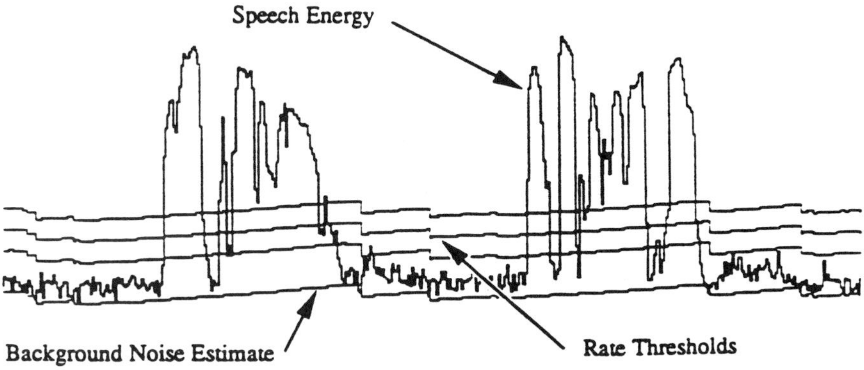

Figure 4: Energy, Background Noise, and Rate Thresholds

Adaptive Rate Decision

QCELP uses an adaptive algorithm to determine the data rate for each frame. The algorithm keeps a running estimate of the background noise energy, and selects the data rate based on the difference between the background noise energy estimate and the current frame's energy.

In each frame, the previous estimate of the background noise energy is compared with the current frame's energy. If the previous estimate is higher than the current frame's energy, the estimate is replaced by that energy. Otherwise, the estimate is increased slightly. Figure 4 shows the energy in a few sentences of speech, and the background noise estimate for these sentences. When no speech is present, the background noise estimate follows the input signal energy. During active speech, the estimate slowly increases, but fluctuations inherent in the energy of the speech signal cause it to be reset continually.

The data rate is then selected based on a set of thresholds which "float" above the background noise estimate, also shown in Figure 4. If the current frame's energy is above all three thresholds, the coder encodes the speech at full rate. If the energy is less than all three thresholds, the coder encodes the speech at eighth rate. If the energy is between the thresholds, the intermediate rates are chosen.[2]

With this algorithm, background noise is almost always coded at eighth rate regardless of its energy. If the background noise suddenly increases, such as when a driver using a car phone opens his window, initially the background noise will be coded at the higher rates. After a few seconds the background noise estimate will rise to the new level of noise and the background will once again be coded at the eighth rate. If the background noise suddenly drops, the estimate immediately drops with it, preventing speech from being coded at the lower rates. In field tests, the algorithm has been shown to be very robust in a

[2]The CDMA system can also force the data rate to be no greater than half rate for certain frames to allow transmission of signaling information in the voice channel (discussed later).

variety of mobile environments.

Error Protection for the CDMA Channel

The QCELP error protection mechanism is built around the CDMA channel. The CDMA system proposed in [1] uses a two frame interleaver to reduce the effects of bursts of errors. A rate 1/2 K=9 convolutional code is used on the forward link, and a rate 1/3 K=9 convolutional code is used on the reverse link. Further details of the system can be found in [1].

For full rate frames, an 11 bit inner CRC is used to protect the 18 most perceptually sensitive bits (the MSBs of the 10 quantized LSPs and the MSBs of the 8 quantized log codebook gains), and a 12 bit outer CRC is used to protect the entire frame, including the inner CRC. For half rate frames, an 8 bit CRC is used to protect the entire frame. All rates have 8 tail bits used to bring the convolutional encoder back to the "all zero" state before the next frame. The CRCs and the tail bits are used at the receiver to determine which data rate was sent by the transmitter.

Undetected bit errors virtually never occur in the CDMA system. The two most common types of channel impairments are "erasures," in which the speech decoder is given no data because the receiver could not determine the transmitted data rate, and "full rate likelies" in which the outer CRC for a full rate frame did not check, but other metrics indicate that the frame was most likely a full rate frame with errors. Since both types of errors are detected errors, the speech decoder can be conservative in reproducing the speech from the corrupted frames of data.

During an "erasure," the LSP frequencies are decayed by 0.9 towards their "white noise" bias values. The previous pitch lag is used, with the pitch gain first saturated at 1 and then decayed towards 0. The decay factor is 0.9, 0.6, 0.3, and 0 for the first, second, third, and fourth erasure in a row. A random codebook vector is chosen and the codebook gain is decayed by 0.7. By decaying the parameters towards their "background noise" values energy is removed from the decoder's filters, creating a slight drop in volume in the reconstructed speech but not an annoying squeak or whistle.

During a "full rate likely" frame, the inner CRC on the most perceptually sensitive bits is used for error detection and correction. If the inner CRC detects no errors or if it indicates that only 1 bit of the 18 most perceptually sensitive bits is in error, the bit in error is corrected and the LSP and codebook data are used as in a full rate frame. The pitch filter is saturated and decayed as in erasures. If the inner CRC shows more than 1 bit in error, the speech decoder treats the frame as an erasure due to the high number of bit errors in the frame.

The erasure rate for full rate frames of the CDMA system operating at capacity is controlled to be between 0.5% and 1.0%. The "full rate likely" rate is typically much less than the erasure rate. At these error rates, the degradation in quality is barely noticeable, and the quality of the speech is very close to that of an error free channel.

ADVANTAGES OF VARIABLE RATE CODING

In addition to increasing system capacity, a variable rate speech coder such as QCELP has many other advantages over both fixed rate transmission schemes and on/off voice activity detection schemes.

Because the low data rates can capture low energy onsets of speech, there is no clipping of the initial parts of words, such as with a voice activity detection system before transmission can be restarted.

Background noise of relatively constant energy is suppressed. Because the low data rates cannot accurately produce the exact, original background noise, the recreated noise generated at these rates typically has a lower volume than the original. In situations with a constant noise in the background, such as when a cellular user is driving a car, the noise is quieted while the speech, coded at full rate, still has the same volume and quality.

While relatively constant background noise is suppressed, interesting short term background noise events are still accurately reproduced without causing a significant decrease in system capacity. Because the coder can select a different data rate every 20 ms, short events in the background, such as a horn honking, can be coded temporarily at the higher data rates, creating a faithful reproduction of the sound. Typically voice activity detection schemes either completely miss these events or encode them at full rate for a fixed period of time, creating a "bursty," artificial sounding background noise.

Signaling information is transmitted in the voice channel with virtually no degradation in speech quality. For signaling information to be sent, the system can instruct the speech coder to encode the speech at half rate for one or two frames, allowing the remaining bits which would have been used for full rate speech data to be used for signaling data instead. The speech is only coded at half rate for 20-40 ms, and the reconstructed speech is indistinguishable from speech coded entirely at full rate. By transmitting speech and signaling data in parallel, the inefficiency of a fixed channel allocation for signaling is eliminated, and overall system capacity and voice quality is increased.

Error protection is matched to the perceptual significance of the data rate and input signal. For example, the QCELP quantizers have reduced dynamic ranges during low rate background noise frames, so any errors which occur in these frames cannot change the coder parameters significantly. Typically errors during low rate frames are perceptually unnoticeable.

Less power is used during lower rate frames, decreasing the average power consumption. For example, in a full duplex implementation of QCELP running on an AT&T DSP 1616 the computation required at full rate, half rate, quarter rate, and eighth rate is 23, 21, 17, and 11 MIPS respectively. Since the DSP can be put in a low power mode when it has finished encoding the current frame and is waiting for the samples for the next frame, the overall power consumption is significantly less than that used by a full rate coder. In addition, in the CDMA system the lower data rates also require less RF transmit power, further lowering the average power consumption.

Variable rate coding allows the CDMA system to have a "soft capacity." In situations where the number of users is greater than the normal system capacity, the average data rate for each user can be decreased slightly by forcing a small percentage of active speech frames to be encoded at half rate rather than full rate. This decreases the interference over the channel and increases the system capacity. Thus, rather than preventing other users from being able to place a call, the system can slightly decrease each user's voice quality while allowing all calls to be placed.

Lastly, because the CDMA system has the capability to transmit at different data rates, it can be easily modified to accept new speech coding technology and data services. For example, if and when half rate coders provide acceptable voice quality, the proposed CDMA system can easily accommodate the new coders, since it can already transmit at half rate. This new system would have roughly twice the capacity of the current CDMA system, or about 30 times analog capacity. Low rate modem or fax services can also be established without any major system redesign.

CONCLUSION

Variable rate speech coding provides many advantages over fixed rate speech coding. QCELP, the first practical variable rate speech coder to be incorporated in a digital cellular system, provides near toll quality speech at an average data rate of under 4 kbit/s, while providing all of the advantages inherent in variable rate coding.

References

[1] QUALCOMM Inc., *Proposed EIA/TIA Interim Standard - Wideband Spread Spectrum Digital Cellular System Dual-Mode Mobile Station - Base Station Compatibility Standard.* Submitted to the TIA TR45.5 Subcommittee, April 21, 1992.

[2] Cellular Telecommunications Industry Association, *Users' Performance Requirements.* September, 1988.

[3] EIA/TIA, *IS-54 Dual-Mode Subscriber Equipment - Network Equipment Compatibility Specification.* 1989.

[4] QUALCOMM Inc., *An Overview of the Application of CDMA to Digital Cellular Systems and Personal Cellular Networks.* Submitted to the TIA TR45.5 Subcommittee, March 28, 1992.

[5] M. R. Schroeder and B. S. Atal, "Code-Excited Linear Prediction (CELP): High Quality Speech at Very Low Bit Rates," in *Proceedings of ICASSP*, 1985.

[6] F. Itakura, "Line Spectrum Representation of Linear Predictive Coefficients of Speech Signals," *J. Acoust. Soc. Amer*, vol. 57, 1975.

12

PERFORMANCE AND OPTIMIZATION OF A GSM HALF RATE CANDIDATE

F.Dervaux, C. Gruet and M. Delprat

MATRA COMMUNICATION
Rue J.P. Timbaud, B.P.26
78 392 Bois d'Arcy Cédex
FRANCE

INTRODUCTION

The "Group Spécial Mobile" (GSM) pan European digital mobile radio system has been designed with a particular TDMA frame structure which enables to use alternatively full rate or half rate channels. Last year, under the control of ETSI, the standardization of a combined speech and channel codec for GSM half rate channels has started. The codec presented by MATRA COMMUNICATION was one of the pre-selected candidates. It has been ranked second in average quality while being the least complex.

To improve its quality and delay performances, the codec was modified twice before the final selection.

I- GSM HALF RATE STANDARDIZATION AND PRE-SELECTED CODEC

The main requirements of GSM half rate channel standardization concern: the global bit rate which has to be 11.4 kbps, the quality which should be equivalent to that of the full rate codec over all transmission conditions, the complexity which has to be limited to four times that of the full rate codec, and the one way transmission delay which should not exceed 90 ms. The pre-selection test selected the 5 best candidates including ours.

Our 6.7 kbps Regular Pulse CELP speech codec[1] [3] [4] has a low complexity thanks to the structure of excitation codebooks which use binary regular pulse sequences. The excitation is the sum of two sequentially determined sequences (Fig. 1): the first with a decimation factor D=4 and the second with a single pulse. Other advantages of this excitation is that no storage is required and that the speech coder is intrinsically robust to transmission errors.

Our codec presented some particular features as compared to others : 30 ms frame length which results in a better efficiency of quantization of Long Term Predictor (LTP) and excitation parameters, and a 16 bits fixed point real time implementation.

The efficient long term prediction comes from the use of fractionnal delays [5] together with a fast closed loop analysis

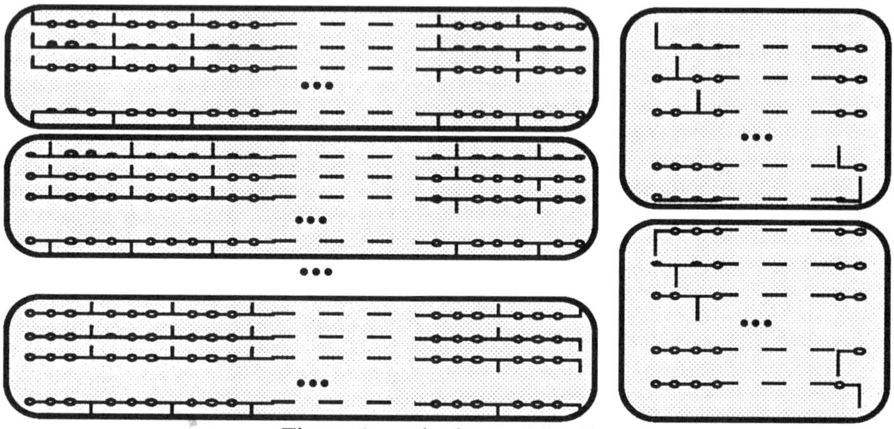

Figure 1: excitation codebooks

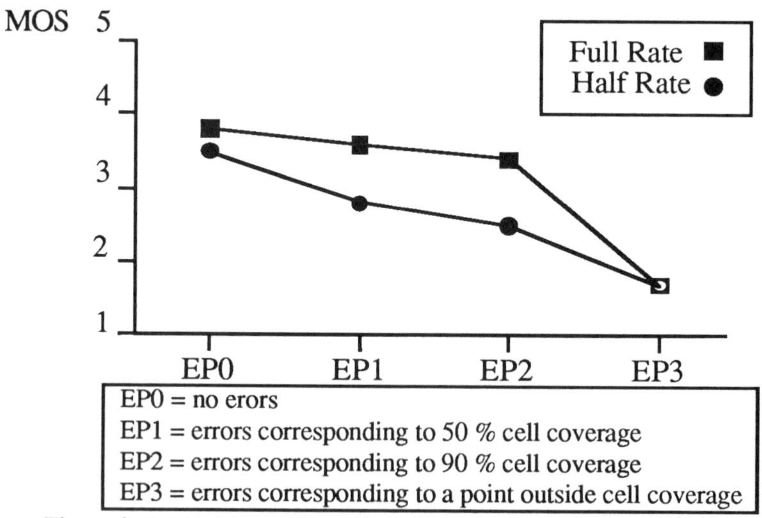

Figure 2: Intrinsic speech quality (Mean Opinion Score (MOS)) for the full rate and the half rate candidate

The channel coder at 4.7 kbps was based on punctured convolutional codes (rate 1/3 and 1/2), error detection (CRC), and interleaving on five bursts. The large interleaving depth partly explains the good resistance to transmission errors but implies a relatively high transmission delay .

Intrinsic speech quality was close to that of the full rate codec though slightly below (around 2dB on average on the equivalent MNRU scale (Fig. 2)). In particular, the sensitivity to input levels and tandeming was not satisfactory. On the other hand, the coder had a very good behaviour in the presence of background noise.

In spite of these promising results, the performances were not sufficient since the objective is to reach the level of qualtity of the full rate codec over all transmission conditions. In addition, for the final selection the delay was strictly limited so that the interleaving depth had to be reduced from 5 to 3. Therefore, the bit rate of the speech coder had to be significantly reduced to allow for a higher redundancy in the channel coding.

II- FIRST MODIFIED SPEECH CODER :5.8 kbps

Roughly, in speech signal, there are two kinds of sounds: the voiced and the unvoiced ones. The adaptive codebook which is the long term predictor contribution is mainly useful for the voiced segments while fixed excitation contains most of the information for the unvoiced ones.

Hence, bit rate reduction could have been efficiently achieved by taking advantage of these specific characteristics. The new speech coder structure uses a voiced/unvoiced decision based on the long term predictor performance in each block of 60 samples. It allows a different coding for voiced and unvoiced blocks, which enables a significant bit rate reduction.

For unvoiced blocks, the adaptative codebook (long term predictor) is supressed and the global excitation is modelled just with one sequence from a binary regular pulse codebook with a decimation factor of 4, plus one sequence from a single pulse codebook with a fixed gain relative to the gain of the first sequence.

For voiced blocks, long term prediction is implemented as it was in the previous codec but the gain quantization has been reduced from 4 to 3 bits and the excitation model uses only the binary regular pulse codebook (the single pulse codebook is supressed).

The voiced/unvoiced decision was based on an estimation of the adaptative codebook performance using the same criterion as for the "excitation gain control", processing described later :

$$Q = \|H.r\|^2 / \|H.e\|^2$$

where H is the weighted impulse response matrix
r is the short term prediction residual signal
e is the residual signal with the contribution of long term prediction subtracted

The block is decided voiced when Q is greater or equal to 2 dB.

The bit allocation is given below (Tab. 1) for this 5.766 kbps speech coder.

The suppression of the single pulse codebook for voiced blocks produces a slight degradation compensated by the "excitation gain control".

	Bits / block		Bits / Frame	
	Voiced	Unvoiced	Voiced	Unvoiced
LP Filter			37	37
Adaptative Codebook				
Index	8		32	
Gain	3		12	
Regular Pulse Codebook				
Index	17	17	68	68
Gain	5	5	20	20
Single Pulse Codebook				
Index		7		28
Voiced / Unvoiced	1	1	4	4
Total	34	30	173	157

Table 1: bit allocation in the first modified codec

A new speech extrapolation procedure is applied whenever the channel decoder indicates a bad frame. Usually, speech substitution is based on a repetition of the parameters of the last valid frames but this does not work well in transitional segments. The explicit voiced/unvoiced decision enables efficient speech substitution because in the case of transition in the last valid frame (i.e. consecutive voiced and unvoiced blocks), substitution is essentially based on the parameters of the last block. It results in a better extrapolation in transitional regions and the quality is clearly improved at high error rates.

On strongly voiced segments, the excitation is often more harmful than helpful to reproduce the harmonic structure of the signal. So we have implemented an "excitation gain control" procedure that applies only on voiced blocks. This procedure is similar to the "constrained excitation" described in [2] but the implementation is different. In fact the original technique produces a significant improvement of the subjective quality for single encoding, but is catastrophic in tandeming conditions if the amount of gain reduction is significant. Thus, in our method the amount of gain reduction is carefully controlled and limited to a reasonable value which also depends on the performance of the adaptative codebook as it is shown in Fig. 3.

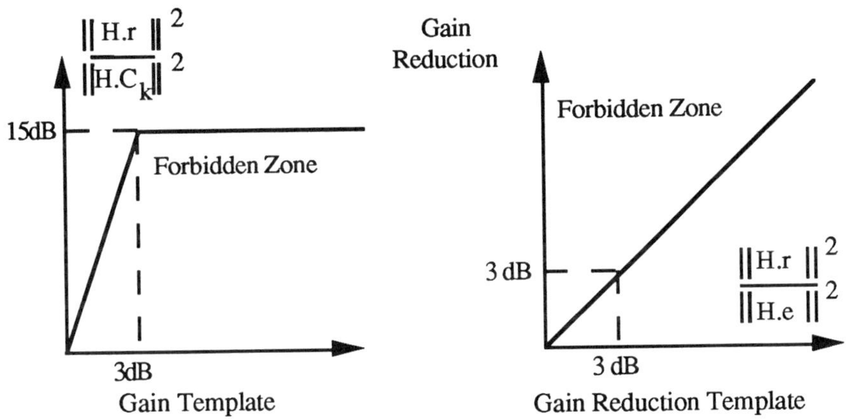

Figure 3: Gain reduction control

With this modified procedure, the quality has been improved for single encoding and also slighty for tandeming.

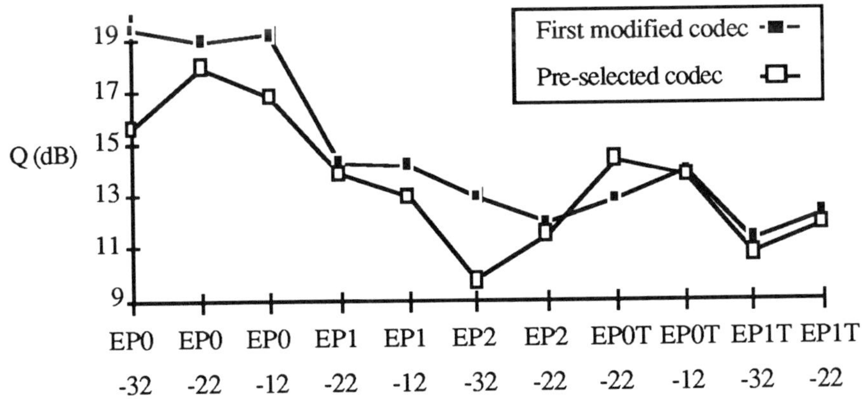

Figure 4: Intrinsic quality over transmission conditions (T appended to EP stands for tandeming conditions)

This new codec has been evaluated over a wide range of conditions [6] including different input levels (-32, -22 and -12db), different error rates (EP0: no error, EP1: C/I=10dB and BER=5%, EP2: C/I=7dB and BER=8%, EP3: C/I=4dB and BER=13%), single encoding or tandeming. Results from a formal subjective test are reported on

Fig. 4 for a subset of representative conditions, in comparison to the performances of the pre-selected codec

It can be seen that intrinsic quality is significantly better for the new codec in spite of the bit rate reduction. In particular, the sensitivity to input levels has been greatly reduced. However, the quality in EP1 and EP0-tandem is still not satisfactory since the quality of the full rate codec remains quite close to the intrinsic quality for these conditions.

III- SECOND MODIFIED SPEECH CODER: 5.0 kbps

To improve further the quality of our speech codec, we decided to allow more flexibility in the fixed excitation model because of the poor statistical properties of our codebook structure.

In [1], we already presented a multi-codebook approach in which the different excitation sequences are sequentially searched. Despite the good quality provided, this method does not efficiently use the available rate.

We designed a better solution where the excitation sequence is selected in a single step using a large codebook composed of several structured codebooks with complementary characteristics. We used four regular pulse (RP) sub-codebooks with different decimation factors (D=8, 12, 15, 60) leading to different pulse densities. Statistics show that the four different sub-codebooks are nearly equally used, which indicates that the global codebook is well designed. The new bit allocation is given in Tab. 2.

	Bits / block	Bits / Frame
LP Filter		37
Adaptative Codebook		
Index	8	32
Gain	3	12
Excitation		
Sub-Codebook	2	8
Index	10	40
Gain	5	20
Total	28	149

Table 2: bit allocation in the second modified codec

This new codec has been compared to the first one within a similar testing framework, to the one descrided below (Fig.5).

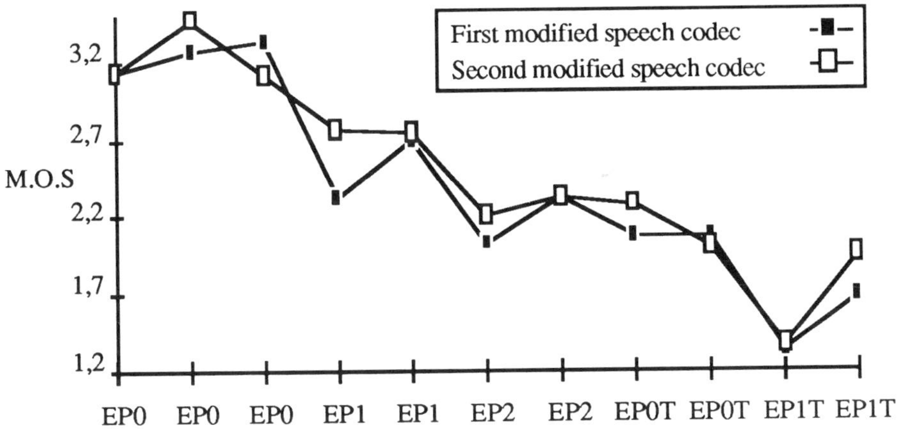

Figure 5: Intrinsic quality over transmission conditions (T appended to EP stands for tandeming conditions)

The second modified codec gives generally the best performances except for EP0 -12dB and has been ranked fourth during the final selection.

CONCLUSION

Thanks to the multicodebook procedure and the modified constrained excitation procedure, intrinsic quality has been improved despite the bit rate reduction. The average quality is improved despite the reduction of interleaving delay. We provided a codec whose complexity and delay are equivallent to that of the full rate codec. However, quality is still not satisfactory in certain conditions such as tandeming.

REFERENCES

[1] C. Gruet, F. Pommaret and M. Delprat, "Experiments with a Regular Pulse CELP Coder for the Pan Europeen Half Rate Channel", in Proceedings ICASSP, 1991.

[2] Y. Shoham, "Constrained-stochastic Excitation coding of speech at 4.8 kb/s", in advances in speech coding, Kluwer Academic Publishers, 1991.

[3] M. Lever and M. Delprat "RPCELP: A high quality and low complexity scheme for narrow band coding of speech", in EUROCON 1988

[4] M.R. Schroeder and B. Atal "Code excited linear prediction (CELP) : High quality speech of very low bit rates", in ICASSP 1985

[5] P.Kroon and Atal "Pitch predictor with high temporal resolution" in ICASSP 1990

[6] H.J. Braun and J.E. Nadvig "European DMR- The standardization procedure on the way from full rate coding to the half rate system", Eurasip workshop Hersbruck September 1989

13

JOINT DESIGN OF MULTI-STAGE VQ CODEBOOKS FOR LSP QUANTIZATION WITH APPLICATIONS TO 4 KBIT/S SPEECH CODING

W.P. LeBlanc[†], S.A. Mahmoud[†], and V. Cuperman[‡]

[†]Systems & Computer Engineering Department,
Carleton University, Ottawa, Canada, K1S-5B6

[‡]School of Engineering Science,
Simon Fraser University, Burnaby, B.C., Canada

INTRODUCTION

Vector Quantization (VQ) of spectral parameters for low-rate speech coding (below 4 kb/s) has recently attracted considerable attention. At this low rate, efficient quantization of the LPC parameters using as few bits as possible is essential. Although spectral parameter quantization was one of the first applications of vector quantization, its use has been limited by concerns regarding computational complexity, lack of robustness, and the expected performance across different speakers, across different spectral shapings, and on noisy communication channels.

The Line Spectrum Pairs [1] (LSPs) are one-to-one transformations of the LPC parameters which result in a set of parameters which can be efficiently quantized while maintaining stability. It is generally accepted that in order to achieve good quality (*transparent*) reconstructed speech [2]:

1. the average spectral distortion should be less than 1 dB, with
2. less than 2 percent outliers having spectral distortion above 2 dB, and
3. no outliers with spectral distortion larger than 4 dB.

In a full search unstructured VQ system, reducing the spectral distortion (SD) to 1 dB requires a large codebook which leads to intractable complexity. By adding suitable structure to the codebook, both the memory and computational complexity can be reduced significantly.

In a multi-stage VQ (MSVQ) system [3,4,5], the parameter vector $\mathbf{x}$ consisting of p (LSP) coefficients is approximated as

$$\begin{aligned}\hat{\mathbf{x}} &= \mathbf{y}_0 + \mathbf{y}_1 + \ldots + \mathbf{y}_{K-1} \\ &= B_0\mathbf{c}_0 + B_1\mathbf{c}_1 + \ldots + B_{K-1}\mathbf{c}_{K-1} \\ &= B\mathbf{c}, \end{aligned} \tag{1}$$

Work partially supported by the Telecommunications Research Institute of Ontario (TRIO) and by the B.C. Science Council through Science and Technology Development Fund.

where $\hat{\mathbf{x}}$ is the quantized version of $\mathbf{x}$, K is the number of stages, and $\mathbf{y}_j$ is the vector selected from the j-th stage. The quantized vector $\hat{\mathbf{x}}$ will also be called the reconstruction vector. The codebook at each stage is comprised of L codevectors. The column vector $\mathbf{c}_j$ consists of the stacked codevectors (j-th stage) and B_j is a sparse Toeplitz matrix (p by Lp) constructed such that $B_j\mathbf{c}_j = \mathbf{y}_j$. Note that B_j is a function of the codebook indices whereas c_j is not. Furthermore $B = [B_0 \quad B_1 \quad \ldots \quad B_{K-1}]$, is a matrix ($p$ by KLp) and $\mathbf{c} = [\mathbf{c}_0^T \quad \mathbf{c}_1^T \quad \ldots \quad \mathbf{c}_{K-1}^T]^T$ is a column vector (KLp by 1). The matrix B is referred to as the selection matrix and B_j is referred to as the selection matrix for the j-th stage. The vector $\mathbf{c}$ is referred to as the stacked codevectors or stacked codebook and the vector $\mathbf{c}_j$ is referred to as the stacked codevectors/codebook for the j-th stage.

A suboptimal sequential search procedure is traditionally used in MSVQ by selecting first the vector $\mathbf{y}_0$ which is *closest* to $\mathbf{x}$, then the vector $\mathbf{y}_1$ which is closest to $\mathbf{x} - \mathbf{y}_0$ and so on. The optimal full search procedure selects the vector $\hat{\mathbf{x}}$ which is closest to $\mathbf{x}$ over the set of possible (L^K) reconstruction vectors

$$\hat{\mathbf{x}} : \left\{ \mathbf{y}_0^{(m)} + \mathbf{y}_1^{(n)} + \ldots \right\} \ \forall \ 0 \le m < L, \ 0 \le n < L, \ \ldots, \tag{2}$$

where $\mathbf{y}_j^{(k)}$ is the k-th codevector from the j-th stage. This is prohibitively complex for values of L and K required to obtain a spectral distortion near 1 dB.

Conventional multi-stage VQ is suboptimal due to the constrained structure of the codebooks, the sequential search procedure, and the stage-by-stage codebook training algorithm.

SEARCH STRATEGY

A weighted mean-square error (WMSE) distortion criterion is used for training the codebooks and for the selection of the best codevectors $\mathbf{y}_j$ at each stage. The WMSE between the original and the quantized parameter vector is defined by

$$d(\mathbf{x}, \hat{\mathbf{x}}) = (\mathbf{x} - \hat{\mathbf{x}})^T W (\mathbf{x} - \hat{\mathbf{x}}), \tag{3}$$

where W is a perceptually based diagonal weighting matrix [6]. This weighting provided slightly lower spectral distortion than the weighting of [2].

The performance of the resulting code is evaluated by the root mean-square spectral distortion between $\mathbf{x}$ and $\hat{\mathbf{x}}$ implemented as in [2]:

$$d_{SD}(A(z), A_p(z)) = \sqrt{\frac{1}{n_1 - n_0} \sum_{n=n_0}^{n_1 - 1} \left[10 \log_{10} \left(\frac{\left| A\left(e^{j2\pi n/N}\right) \right|^2}{\left| A_p\left(e^{j2\pi n/N}\right) \right|^2} \right) \right]^2} \tag{4}$$

where n_0 and n_1 correspond to 125 Hz and 3.1 kHz, respectively, and $A(z)$ and $A_p(z)$ represent the quantized and unquantized model filters respectively. In practice, $n_0 = 4$, $n_1 = 100$, and an $N = 256$-point FFT was used to compute $A(e^{j2\pi n/N})$ and $A_p(e^{j2\pi n/N})$.

The performance of a multi-stage VQ can be improved by using an M-L tree search procedure [3,5], rather than the conventional sequential search procedure described in the introduction. In the sequential search procedure only a single best set of indices is maintained from one stage to the next, whereas in the M-L procedure M best sets of indices are maintained. The M-L procedure provides a good trade-off between the poor performance of a sequential search procedure and the large computational complexity of the full search procedure.

It was determined experimentally that the M-L search applied to multi-stage VQ achieves performance very close to that of the optimal search for relatively small values of M ($M \leq 16$) [3,4].

CODEBOOK DESIGN

Traditionally, multi-stage VQ codebooks are trained sequentially, each stage using a training sequence consisting of quantization error vectors from the previous stage. This design approach is clearly suboptimal since all stages are not included in the minimization procedure. In this section, a new design procedure is introduced using a generalized Lloyd algorithm to minimize average WMSE based on a training sequence.

Simultaneous Joint Codebook Design

The goal of simultaneous joint codebook design is to jointly optimize all codevectors over all stages after each iteration, and therefore joint design can converge far quicker, and to possibly a lower final distortion. The concept is to optimize the set of possible reconstruction vectors defined in (2), rather than minimizing the WMSE after each stage.

Let $\{\mathbf{x}(n)\}$ be a training sequence of spectral vectors and $\hat{\mathbf{x}}(n)$ the reconstruction vector corresponding to the source vector $\mathbf{x}(n)$, where $\hat{\mathbf{x}}(n) = B(n)\mathbf{c}$. Note that $B(n) = [\, B_0(n) \quad B_1(n) \quad \ldots \quad B_{K-1}(n)]$ is the selection matrix corresponding to the source vector $\mathbf{x}(n)$. With this notation, the average distortion over the training sequence at iteration r, d_r, can be written as

$$\begin{aligned} d_r &= \sum_n (\mathbf{x}(n) - \hat{\mathbf{x}}(n))^T W(n)(\mathbf{x}(n) - \hat{\mathbf{x}}(n)) \qquad (5) \\ &= \sum_n \mathbf{x}^T(n)W(n)\mathbf{x}(n) - 2\mathbf{c}^T \sum_n B^T(n)W(n)\mathbf{x}(n) + \mathbf{c}^T \sum_n B^T(n)W(n)B(n)\mathbf{c}, \end{aligned}$$

which leads to

$$d_r = d_0 - 2\mathbf{c}^T\mathbf{y} + \mathbf{c}^T Q\mathbf{c}. \qquad (6)$$

As with the generalized Lloyd algorithm, the selection matrices $\{B(n), \forall\, n\}$ are determined given $\mathbf{c}$, and $\mathbf{c}$ is then computed for the given selection matrices. The minimizing solution satisfies $Q\mathbf{c} = \mathbf{y}$. It is tempting to write the minimizing solution of (6) in the form $\mathbf{c} = Q^{-1}\mathbf{y}$. However, in general, the inverse does not exist, since Q is not full rank. An infinite number of solutions for the joint codebook exist since adding a constant vector $\mathbf{v}$ to all the vectors of any one stage, while subtracting the same constant vector from all the vectors of any other stage leads to the same set of possible reconstruction vectors $\hat{\mathbf{x}}$ for any $\mathbf{v}$. All minimizing solutions of (6) are equivalent in terms of the set of reconstruction vectors each can generate. A number of techniques such as Newton's method, steepest descent, or conjugate gradient can be used to determine a solution which minimizes (6) and thus satisfies $Q\mathbf{c} = \mathbf{y}$.

A projection method (to minimize (6)) is used here for simplicity reasons. The stacked codebook for each stage is optimized sequentially during the same iteration r, by holding all $\mathbf{c}_k$, $k \neq j$ fixed in (6), for $j = 0, 1, \ldots, K-1$. The stacked codebook $\mathbf{c}$ is written as a function of the stacked codebooks for each stage

$$\begin{aligned} \mathbf{c} &= [\,\mathbf{c}_0^T \quad \mathbf{c}_1^T \quad \ldots \quad \mathbf{c}_{j-1}^T \quad \mathbf{c}_j^T \quad \mathbf{c}_{j+1}^T \quad \ldots \quad \mathbf{c}_{K-1}^T\,]^T \\ &= \bar{\mathbf{c}}_j + S_j \mathbf{c}_j, \end{aligned} \tag{7}$$

where S_j is a simple shifting matrix and

$$\bar{\mathbf{c}}_j = [\,\mathbf{c}_0^T \quad \mathbf{c}_1^T \quad \ldots \quad \mathbf{c}_{j-1}^T \quad \mathbf{0}_{pL} \quad \mathbf{c}_{j+1}^T \quad \ldots \quad \mathbf{c}_{K-1}^T\,], \tag{8}$$

where $\mathbf{0}_{pL}$ is an all-zero pL-dimensional vector. Substitution of (7) into (6) leads to

$$d_r = d_{rj} - 2\mathbf{c}_j^T \mathbf{y}_j + \mathbf{c}_j^T Q_{jj} \mathbf{c}_j, \tag{9}$$

where $d_{rj} = d_0 - 2\bar{\mathbf{c}}_j^T \mathbf{y} + \bar{\mathbf{c}}_j^T Q \bar{\mathbf{c}}_j$, $Q_{jj} = S_j^T Q S_j$ and $\mathbf{y}_j = S_j^T (\mathbf{y} - Q\bar{\mathbf{c}}_j)$. Thus, the j-th stage stacked codebook is given by

$$\mathbf{c}_j = Q_{jj}^{-1} \mathbf{y}_j. \tag{10}$$

Conveniently, Q_{jj} is diagonal (since $W(n)$ is), which leads to a relatively simple evaluation of $\mathbf{c}_j$.

The vector $\mathbf{y}_j$ is a function of $\mathbf{c}$, Q and $\mathbf{y}$. Thus the full vector $\mathbf{y}$ and matrix Q must be stored during the design procedure. Clearly, $\mathbf{c}_j$ in (10) depends on the stacked codebook for all other stages. For a given partition of the training sequence, the algorithm repeatedly re-designs the stacked codebooks $\mathbf{c}_j$, $j = 0, 1, \ldots K-1$ until convergence is reached. This inner loop provides a re-optimized stacked codebook $\mathbf{c}$.

Once $\mathbf{c}$ has been re-optimized, the training sequence is again repartitioned given the new codebooks. This method is referred to as simultaneous joint design since all codebooks are re-optimized simultaneously and jointly after each pass over the training sequence.

In the joint design algorithm, the codebook optimization is done under the assumption that a full search is used. If a sequential search or a tree search are used, the performance of the code may degrade. Theoretically, joint design combined with sequential search may even result in worse performance than sequentially designed codebooks. However, the tree search approaches the full search performance for a moderate value of M [3,4] and therefore the jointly designed codebooks are expected to perform well with a tree search with large M. Moreover, a relatively simple empirical procedure when applied to the joint design algorithm was found to result in *robust* codebooks which have good performance for sequential and tree search while not affecting the set of reconstruction values. This empirical procedure is shown as step 5 of the joint design procedure presented in Fig. 1.

Monotonic convergence to a local minimum of the multidimensional distortion function is guaranteed if a full search procedure is used. Detailed convergence properties of the design algorithms is beyond the scope of this paper and the interested reader is referred to [7].

1. *Initialize.* Set $r = 1$. Create K stage random codebook.
2. *Partition.* Set all elements of $\mathbf{y}$ and Q to zero, then for each vector in the training sequence, determine the codebook indices which minimize the WMSE. Compute both $B^T(n)W(n)B(n)$ and $B^T(n)W(n)\mathbf{x}(n)$ and update the running sum Q, $\mathbf{y}$ and d_r.
3. *Compute new stacked codebook.* For $j = 0$ to $K-1$ compute $\mathbf{c}_j = Q_{j,j}^{-1}\mathbf{y}_j$ where $Q_{j,j} = S_j^T Q S_j$ and $\mathbf{y}_j = S_j^T(\mathbf{y} - Q\bar{\mathbf{c}}_j)$.
4. *Convergence of* $\mathbf{c}$. Repeat step 3 until convergence.
5. *Reorder.* a) Ensure that the sum of all codevectors for each stage (except the first stage) is the all zero vector

$$\sum_{k=0}^{L-1} \mathbf{y}_j^{(k)} = \mathbf{0}_p \quad \forall \ j > 0,$$

where $\mathbf{y}_j^{(k)}$ is the k-th codevector from the j-th stage and $\mathbf{0}_p$ is an all-zero p-dimensional vector. b) Switch the order of the codebooks such that the energy in $\mathbf{c}_j$ is less than the energy in $\mathbf{c}_k$, for all $j > k$

$$\mathbf{c}_j^T \mathbf{c}_j \leq \mathbf{c}_k^T \mathbf{c}_k \quad \forall \ j > k.$$

(The energy of the first codebook is computed after the mean is subtracted).
6. *Convergence test.* If $|d_{r-1} - d_r|/d_r > \epsilon$, set $r = r + 1$, and goto 2.
7. *Terminate.*

Fig 1: MSVQ Simultaneous Joint Design Algorithm

Outlier Weighting

One of the problems of concern in VQ of the LPC parameters are the so-called outliers – input vectors which are poorly represented in the training sequence and are quantized with a spectral distortion much larger than the average. The outlier performance of a VQ can be significantly improved by appropriately weighting the distortion measure during the centroid computation. The training sequence is (as before) partitioned according to the nearest neighbour criterion by minimizing (5) and in the centroid calculation a weighted error is minimized according to

$$d_r = \sum_n f(SD)(\mathbf{x}(n) - \hat{\mathbf{x}}(n))^T W(n)(\mathbf{x}(n) - \hat{\mathbf{x}}(n)), \tag{13}$$

where $f(SD)$ is some (scalar) function of the codebook at iteration r and the spectral distortion (SD) between $\mathbf{x}(n)$ and $\hat{\mathbf{x}}(n)$ (in dB). The outlier weighting function ($f(SD)$) is only used in the centroid computation. Although convergence is not guaranteed since the centroid computation and the codebook search criteria are different, the algorithm was observed to converge in practice [4,7]. A number of functions were investigated, and it was found that $f(SD) = SD^\rho$ ($\rho \geq 0$) resulted in a good trade-off between outliers and average spectral distortion.

PERFORMANCE AND COMPLEXITY TRADE-OFFS

In sequentially searched MSVQ, the design has been oriented toward the largest implementable codebooks and the smallest number of stages. For example, a quantizer using 24 bits per frame would typically be implemented using two codebooks of size 4096 (4096-2). Increasing the number of stages for sequentially searched MSVQ leads to a quick degradation in performance. The introduction of tree search for multi-stage VQ leads to a significant improvement in the performance, particularly for configurations having a relatively large number of small codebooks.

The search complexity is defined as the number of arithmetic operations required to obtain the quantized vector and is presented on a logarithmic scale [4]. For tree-searched MSVQ the complexity is given by

$$C_{\mathrm{MSVQ}} = \log_2 \left(3pL + \sum_{j=1}^{K-1} \min\left(L, \min\left(M, L^j\right)\right)p + \left(\min\left(M, L^j\right) + 2\right)Lp \right),$$

and for split VQ [2], the complexity is $C_S = \log_2(3L(n_0 + n_1 + \ldots + n_{K-1}))$, where n_j is the dimension of the j-th codebook. These relations will be used for evaluating the complexity in the experimental results presented below.

The downsampled TIMIT-TRAIN database (SX sentences) was used for training the codes in this section. The autocorrelation method of LPC analysis was used on 160 sample frames with 16 samples of overlap on each of the previous and next frames. High frequency correction (as in [8]) was applied to the system of linear equations and 10 Hz of bandwidth expansion was applied after solving the linear system of equations. The TIMIT-TRAIN database consisted of 339,850 vectors. The codes were tested on the TIMIT-TEST database, processed in the same manner as the TIMIT-TRAIN database. The TIMIT-TEST database consisted of 121,200 vectors.

The performance (spectral distortion and outliers between 2 and 4 dB) of a 16-6 code (16 levels per stage, 6 stages) utilizing $M = 4$ and joint design (24 iterations) for various values of ρ is shown in Table 1. The various values of ρ provide a trade-off between outlier performance and average spectral distortion. Weighting with $\rho = 2$ was used during the design procedure for the codes described in the rest of the paper.

Figure 2a shows average spectral distortion versus complexity for four tree-searched MSVQ configurations and one split VQ configuration, all operating at 24

Table 1: Spectral Distortion and Outliers for Various Weightings (both inside and outside the training sequence).

Weighting	SD (dB)		Outliers (2-4 dB) %	
	Inside	Outside	Inside	Outside
$\rho = 0$	1.11	1.13	2.04	2.32
$\rho = 1$	1.12	1.13	1.46	1.84
$\rho = 2$	1.12	1.14	1.18	1.42
$\rho = 5$	1.17	1.19	0.78	1.20

bits/frame. The 4096-2 split VQ code used the same partitioning of the LSP vector as in [2]. Figure 2a shows that although the 4096-2 multi-stage code achieves the lowest spectral distortion, better complexity-distortion trade-offs may be obtained using codes having more than two stages. For example, a spectral distortion lower than 1 dB can be obtained at much lower complexity using the 64-4 code with $M = 8$ than using the 4096-2 code. Figure 2a shows that a range of complexity-distortion trade-offs can be obtained for each multi-stage code by selecting a suitable value for M.

Figure 2b shows average spectral distortion versus complexity at rates of 22-30 bits/frame. At a rate of 22 bits/frame the 2048-2 code (with $M = 2$) obtains performance virtually identical to the 24 bits/frame (4096-2) split VQ code. Figure 2b shows that by increasing the rate to 24 bits/frame (64-4 code), 28 bits/frame (16-7 code), or 30 bits/frame (4-15 code) the performance may be improved while decreasing the computational complexity. Again, for a particular code, trade-offs between complexity and performance can be made by selecting a suitable value of M. In all configurations of Fig. 2 (with M chosen such that $SD \leq 1$ dB), there were no outliers larger than 4 dB and the percentage of outliers larger than 2 dB was under 1%.

The complexity and rate required to obtain near 1 dB average spectral distortion for various codes is displayed in Table 2. A spectral distortion of 1 dB can be obtained by multi-stage VQ with only two codevectors per stage and 28 stages.

One of the best configurations in terms of the trade-off between complexity and performance for 4 kb/s speech coding in Fig. 2a and Table 2 is 64-4 which achieves

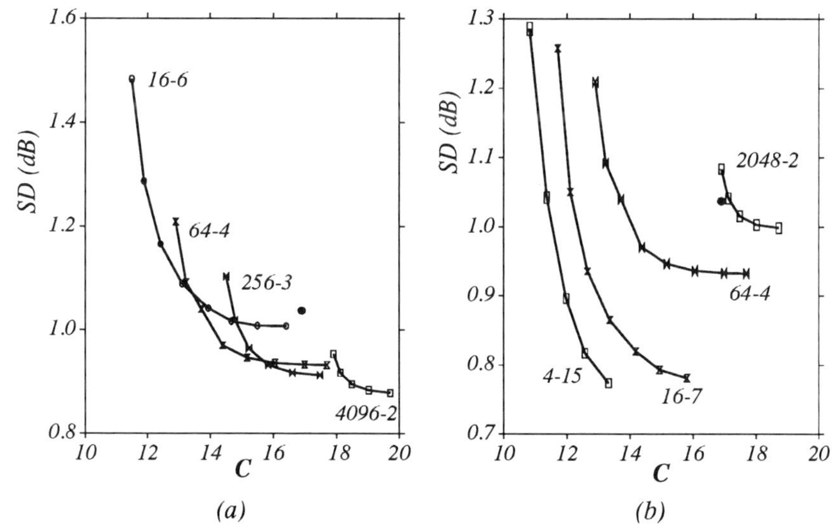

Fig 2: Average Spectral Distortion Performance of Tree-Searched MSVQ versus Complexity. (a) Rate of 24 bits/frame. (b) Rates of 22-30 bits/frame. The solid points in both (a) and (b) are for a split VQ with L=4096, and K=2. The codes are referred to as L-K, where L is the number of levels per stage, and K is the number of stages. The successive points on each curve correspond to M=1, 2, 4, 8, ...

Table 2: Different Tree-Searched MSVQ codes to obtain approximately equivalent average spectral distortion. The rate is given in bits/frame.

Code	M	C	Rate	SD (dB)	% Outliers 2-4 dB	% Outliers > 4 dB
S-4096-2	-	16.9	24	1.04	0.53	0.00
2048-2	2	17.1	22	1.04	0.67	0.00
64-4	4	13.7	24	1.04	0.47	0.00
16-6	16	13.9	24	1.04	0.59	0.00
4-13	8	12.4	26	1.03	1.49	0.01
16-7	2	12.1	28	1.05	0.80	0.00
2-28	8	11.5	28	1.00	0.48	0.00

a spectral distortion of about 1 dB at a complexity more than 8 times lower than the split VQ (4096-2) code. Moreover, 64-4 requires storing only 256 codevectors as compared to 8192 codevectors required by 4096-2. Note that the 28 bits/frame system (16-7) has a very low computational complexity at an average spectral distortion of 1 dB and a memory complexity of only 112 codevectors.

MULTI-LANGUAGE AND INPUT RELATED ROBUSTNESS

One of the potential problems in using vector quantization for low-rate speech coding is the lack of robustness across different languages and different input processing techniques. An example of different input processing techniques are the IRS spectral weighting typical of telephone speech and the flat spectral shaping characteristic of high quality microphones. This section presents results obtained by multi-stage codes trained using the English TIMIT-TRAIN data base when tested on data-bases in different languages using different input spectral shapings.

Table 3 shows the spectral distortion and outlier performance of tree-searched MSVQ for (a) German (2,297 vectors), (b) Italian (2,333 vectors), and (c) Norwegian (1,416 vectors) speech data bases. The foreign language database includes IRS weighted speech which was used for testing codecs in the CCITT 16 kb/s low-delay competition. Note the good robustness across languages for all tree-searched MSVQ systems tested.

In the same Table, (e) displays the performance on the TIMIT-TEST database (121,200 vectors), while (d) shows the performance on an English test database consisting of speech recorded through a high quality microphone (28,000 vectors). The IRS weighted databases and the TIMIT databases have similar average spectral characteristics (spectral roll off of approximately 2 dB/octave) whereas the English database has a somewhat higher spectral roll off (approximately 5 dB/octave). For these cases, the higher rate systems having a large number of small stage codebooks (such as 8-9) show significantly better robustness than the lower rate larger codebook 4096-2 systems (including split VQ). Although similar performance was observed both inside the training sequence and on the TIMIT-TEST sequence, on foreign

Table 3: Spectral Distortion and Outlier Performance on Different Language and Input Spectral Shapings. (a) German (b) Italian (c) Norwegian (d) English (e) TIMIT-test.

		Average SD (dB)					% Outliers (2-4 dB)				
Code	M	(a)	(b)	(c)	(d)	(e)	(a)	(b)	(c)	(d)	(e)
64-4	4	1.13	1.10	1.12	1.22	1.04	1.48	1.42	1.63	5.16	0.47
16-6	16	1.13	1.10	1.05	1.19	1.04	1.83	1.80	2.2	3.86	0.59
S-4096-2	-	1.13	1.08	1.10	1.20	1.04	1.40	0.56	1.70	2.69	0.53
8-9	4	1.10	1.08	1.02	1.12	1.04	0.96	1.24	1.70	2.52	0.54
2-27	16	1.03	1.00	0.97	1.00	1.02	0.87	0.43	1.63	1.29	0.61

language databases and on databases with different spectral shapings the codes with a large number of stages are more robust, and have very low complexity.

The results presented above show that robust VQ can be accomplished by using multi-stage codes with a relatively large number of stages. Increasing the number of stages adds structure to the code and results in increased robustness at the expense of a small degradation in average spectral distortion.

REFERENCES

[1] P. Kabal and R. Ramachandran, "The Computation of Line Spectral Frequencies Using Chebyshev Polynomials," *IEEE Trans. on ASSP*, vol. ASSP-34, Dec. 1986.

[2] K. Paliwal and B. S. Atal, "Efficient Vector Quantization of LPC Parameters at 24 bits/frame," *ICASSP*, pp. 661–664, March 1991.

[3] B. Bhattacharya, W. P. LeBlanc, S. A. Mahmoud, and V. Cuperman, "Tree Searched Multi-Stage Vector Quantization of LPC Parameters For 4 kb/s Speech Coding," *ICASSP*, pp. 105–108, May 1992.

[4] W. LeBlanc, V. Cuperman, B. Bhattacharya, and S. A. Mahmoud, "Efficient Search and Design Procedures for Robust Multi-Stage VQ of LPC Parameters for 4 kb/s Speech Coding," *Submitted to IEEE Trans. on ASSP*, May 1992.

[5] N. Phamdo, N. Favardin, and T. Moriya, "Combined Source-Channel Coding of LSP Parameters Using Multi-Stage Vector Quantization," *IEEE Workshop on Speech Coding for Telecommunications*, pp. 36–38, 1991.

[6] F. F. Tzeng, "Analysis-By-Synthesis Linear Predictive Speech Coding at 2.4 kbit/s," *Proc. Globecom 89*, pp. 1253–1257, 1989.

[7] W. P. LeBlanc, *CELP Speech Coding at Low to Medium Bit Rates*. PhD thesis, Carleton University, 1992.

[8] B. S. Atal and M. R. Schroeder, "Predictive Coding of Speech Signals and Subjective Error Criteria," *IEEE Transactions on Acoustics Speech and Signal Processing*, vol. ASSP-27, pp. 247–254, June 1979.

14

WAVEFORM INTERPOLATION IN SPEECH CODING

W. Bastiaan Kleijn

Speech Research Department
AT&T Bell Laboratories
Murray Hill, NJ 07974, USA

Wolfgang Granzow

Philips Kommunikations Industrie
Thurn-und-Taxis-Str. 10
W-8500 Nuernberg 10, Germany

INTRODUCTION

In waveform coders, the quantized values of the transmitted parameters are selected on the basis of a fidelity criterion comparing the original and reconstructed speech signals. An important class of waveform coders is formed by the analysis-by-synthesis coders [1], which include code-excited linear prediction (CELP). In these coders, a multitude of trial reconstructed signals is generated for a large selection of quantization levels of the coder parameters. The fidelity criterion is then used to select a good set of quantization levels for the parameters.

The advantage of waveform coders is that, in a proper setup, the reconstructed speech signal converges to the original signal with increasing bit rate. Thus, an increased bit rate can compensate for deficiencies in the model used to describe the speech signal. Generally, the fidelity criterion is a least mean-square error criterion operating on the spectrally-weighted original and reconstructed signals. The spectral weighting accounts for the spectral masking of the human auditory system [2].

A waveform-matching procedure implicitly places onto the reconstructed speech constraints which are not required for good speech quality. Relaxation of these constraints results in a decrease in bit rate while good speech quality is maintained [3]. The pitch is a good example of a parameter which requires a high bit rate as a result of the waveform-matching procedure. The error criterion has resulted in updates of the pitch values every 2.5-7.5 ms in most current analysis-by-synthesis coders. However, relatively large deviations from the original pitch contour do not affect the perceived speech quality as long as the smoothness of the original contour is maintained.

Another example of the strict constraints which waveform-matching imposes results from the interaction of the waveform shape and the periodicity. Accurate preservation of the level of periodicity of the speech signal is imperative for good quality. To obtain this high accuracy over the entire signal bandwidth in a conventional waveform-matching procedure, high accuracy of the waveform shape (and thus a high bit rate) is required.

Recognition that voiced speech can be modeled as a concatenation of slowly evolving pitch-cycle waveforms with an added noise signal leads to a relaxation of the waveform-matching constraints. The noiseless signal can be described as a sequence of *prototype* waveforms, updated at regularly or irregularly spaced time instants. If these time-instants are sufficiently close (usually 20-30 ms), the intermediate pitch-cycle waveforms can be approximated by interpolation of the two nearest prototype waveforms. A reconstructed speech signal can be obtained by concatenation of these interpolated pitch-cycle waveforms and adding an appropriate noise signal. In this prototype-waveform interpolation (PWI) approach, waveform matching is performed on the prototype waveforms instead of on the entire speech signal. Thus, the PWI coder is not constrained to reproduce the original pitch contour accurately, and the level of periodicity is independent of the waveform-matching accuracy.

In the present paper we discuss a blockwise implementation of the PWI coder. For a discussion of other PWI and related algorithms we refer to [3-8]. We use the PWI method in conjunction with linear prediction (LP) methods. Standard methods exist for quantization of the LP description of the spectral envelope, and the associated residual prototype waveform can be quantized using the analysis-by-synthesis procedures familiar from CELP. Discontinuities, which may be present at the pitch-cycle boundaries, are rendered inaudible if the concatenation of prototype waveforms is performed in the residual domain. A final advantage of performing PWI in the LP-residual domain is that most of the perceptually significant information of the residual signal is located near the pitch pulses, making the choice of prototype boundaries less critical.

We now proceed with a section on the blockwise PWI method, followed by a section where experimental results are discussed. We end with a conclusion section.

BLOCKWISE PROTOTYPE-WAVEFORM INTERPOLATION

First a prototype waveform representative of the original signal near the update time instant (the future-side boundary of the current update frame) must be extracted. It is efficient to extract the prototype waveform from the upsampled residual signal, using a pitch-period estimate as an aid. The pitch period can be obtained from a standard procedure [9]. A time interval (e.g. 25 ms) is defined centered around the update time instant. The maximum absolute value of the upsampled residual signal within this interval is located. This is a first pitch pulse location. Then, a recursive search for more pitch pulses is performed by searching for absolute maxima at a distance of approximately one pitch period from the known pitch pulses. Pitch-pulse markers found according to this procedure are shown in Figure 1b. The time location t_m of the pitch pulse nearest to the update instant is identified in this manner and used as the center for the prototype waveform. The unquantized prototype excitation waveform is obtained by applying a rectangular window of length one pitch period to the residual signal. If $e(t)$ is the residual signal, $p(t)$ is the pitch period, and $\Xi(t,a)$ is a rectangular (boxcar) window of length a centered at the origin then

$$\tilde{u}_m(\tau) = e(\tau + t_m)\, \Xi(\tau, p(t_m)) \tag{1}$$

is the unquantized prototype excitation waveform. (In this paper, we will denote the various signals as continuous functions of time, in a digital implementation the operations are performed on the upsampled signals.) This extraction procedure works well with the blockwise interpolation method described below because the boundaries of the prototype waveforms are generally located in areas of low energy.

Prior to quantization, the present prototype excitation waveform must be aligned with the previous, quantized prototype excitation waveform. Let us write the previous, quantized waveform as $u_{m-1}^{(Q)}(\tau, \boldsymbol{i}_{m-1})$, where the vector $\boldsymbol{i}_{m-1}$ describes the quantization indices for the codebook and gain indices. Then we align the prototypes according to:

$$u_m(\tau) = \tilde{u}_m(\tau - \xi), \tag{2}$$

where the alignment shift is:

$$\xi = \underset{\xi'}{argmin}\, \mathrm{D}(\, u_{m-1}^{(Q)}(\tau, \boldsymbol{i}_{m-1}),\, \tilde{u}_m(\tau - \xi')). \tag{3}$$

In (3), $D(.,.)$ can be a simple least-squares error criterion or a cross correlation operating directly on the prototype excitation waveforms. The alignment (3) implies that the main pulse as defined by the pitch marker will be displaced from the origin of the prototype waveform. To prevent a drift of this main pulse location over updates, it is important to align the past quantized prototype waveform with a single, centered pulse prior to the alignment operation (3). From here on, all $u_{m-1}^{(Q)}(\tau, \boldsymbol{i}_{m-1})$ are assumed to have been aligned in this manner. Keeping the main pitch pulse centered is also beneficial if trained codebooks are used for encoding the prototype waveform.

After alignment, differential quantization can be applied. Let $H[.]$ denote a filtering operation which adds a perceptually relevant spectral weighting, in a fashion similar to the spectral weighting in CELP. Furthermore, let $SNR\,(w_1(\tau), w_2(\tau))$ denote the signal-to-noise ratio between the signal waveform $w_1(\tau)$ and the (quantization) noise waveform $w_2(\tau) - w_1(\tau)$. Then the quantization procedure is given by:

$$u_m^{(Q)}(\tau, \boldsymbol{i}_m):$$

$$SNR(H[u_m(\tau)],\, H[u_m^{(Q)}(\tau, \boldsymbol{i}_m)]) = \min_{\boldsymbol{i}'_m} SNR(\, H[u_m(\tau)],\, H[u_m^{(Q)}(\tau, \boldsymbol{i}'_m)]). \tag{4}$$

By transmitting the vector of quantization indices $\boldsymbol{i}_m$ and an index to the pitch period, the prototype excitation waveform can be reconstructed at the receiver.

Although the quantization (4) works satisfactorily at high bit rates, it gives rise to reverberation at lower bit rates. This reverberation results from larger fluctuations in the waveform shape in the reconstructed signal than in the original signal. These fluctuations can be quantified in terms of the normalized cross correlation between the unquantized and quantized prototype waveforms, after optimal alignment, or the signal-to-change ratio (SCR) [3,4] between the prototype waveforms. The SCR between two prototype excitation waveforms $w_1(\tau)$ and $w_2(\tau)$ is defined as:

$$SCR(w_1(\tau), w_2(\tau)) =$$

$$\max_{\phi} \left(1 - \frac{\left(\int_{-\infty}^{\infty} H[w_1(\tau)] H[w_2(\tau - \phi)] d\tau\right)^2}{\int_{-\infty}^{\infty} H[w_1(\tau)] H[w_1(\tau)] d\tau \int_{-\infty}^{\infty} H[w_2(\tau)] H[w_2(\tau)] d\tau}\right)^{-1}. \tag{5}$$

The advantage of using the SCR is that its values are on a similar scale as the SNR values obtained in the quantization (they are conveniently expressed in dB). To prevent increased fluctuations, the quantization procedure of (4) must be performed under the following constraint;

$$SCR(u_m^{(Q)}(\tau, \boldsymbol{i}_m),\, u_{m-1}^{(Q)}(\tau, \boldsymbol{i}_{m-1})) = SCR(u_m(\tau),\, u_{m-1}(\tau)). \tag{6}$$

To satisfy this constraint, it is convenient to orthogonalize the codebook entries to the previous quantized prototype waveform (in the spectrally-weighted domain). Then, the SCR depends solely on the gains of $u_{m-1}^{(Q)}(\tau, \boldsymbol{i}_{m-1})$ and the selected codebook entries. Thus, the shape can be quantized first, and then the optimal gains can be determined under the constraint (6).

Generally, the constraint (6) on the quantization (4) means that a decrease in the SNR of the quantized prototype waveforms is traded for an increase in the SCR. Despite the decrease in the SNR, this is associated with an increase in speech quality resulting from the removal of reverberation.

At the receiver the following interpolation algorithm can be used to obtain the excitation waveform from the prototype excitation waveforms. We denote by t_k the center of pitch cycle number k within the current interpolation interval, and by p_k the pitch period of pitch cycle k. It is convenient to number the pitch periods in an interpolation interval from k=0 to k=K. We interpolate the pitch period linearly with the pitch-cycle index k from the transmitted values:

$$p_k = \frac{K-k}{K}\, \mathrm{p}(\mathrm{t}_{m-1}) + \frac{k}{K}\, \mathrm{p}(\mathrm{t}_m), \quad k = 0, 1, \cdots, K. \tag{7}$$

The time locations, t_k, of the pitch-cycle centers at the receiver are obtained by adding the pitch periods, or, equivalently:

$$t_k = t_0 + \frac{k}{2}\,(p_0 + p_k). \tag{8}$$

To enhance performance (and lessen the impact of channel errors) it is advantageous to align $u_m^{(Q)}(\tau, \boldsymbol{i}_m)$ with $u_{m-1}^{(Q)}(\tau, \boldsymbol{i}_{m-1})$ prior to interpolation, using the procedure of (3). We denote the resulting alignment shift by ζ. The excitation pitch-cycle waveforms $v(0, \tau), \cdots, v(K, \tau)$ are obtained from linear interpolation with the pitch-cycle index:

$$v(k, \tau) = \frac{K-k}{K}\, u_{m-1}^{(Q)}(\tau, \boldsymbol{i}_{m-1}) + \frac{k}{K}\, u_m^{(Q)}(\tau-\zeta, \boldsymbol{i}_m), \quad k = 0, 1, \cdots, K. \tag{9}$$

The excitation waveform $x(t)$ is obtained by concatenation of the truncated waveforms, starting at t_0. Using the same rectangular window function as used in (1):

$$x(t) = \sum_{k=0}^{K} v(k, t - t_k)\, \Xi(t - t_k - \frac{k}{K}\zeta,\, p_k + \frac{\zeta}{K}), \qquad t_0 \le t < t_K + \zeta. \tag{10}$$

The receiver performs the following processing. To start, it receives over the channel the encoded pitch periods p_0 and p_K, and the encoded excitation waveforms $u_{m-1}^{(Q)}(\tau, \boldsymbol{i}_{m-1})$ and $u_m^{(Q)}(\tau, \boldsymbol{i}_m)$. From the previous interval t_0 is known, and the end of the update frame is the desired endpoint of the update interval t_{update}. By substituting t_{update} for t_k in equation (8), a noninteger value is obtained for k, which is rounded up to the next integer, defined as K. Then the actual t_K is computed using equation (8).

Next, the t_k for the intermediate pitch are computed using (7) and (8). The quantized prototypes are aligned according to the procedure of (3). The intermediate pitch-cycle waveforms are computed using (9) and the excitation signal is computed using (10). The value of t_0 for the next interpolation interval is set to $t_K+\zeta$. Finally, filtering $x(t)$ with an LP-synthesis filter results in the reconstructed speech signal.

In the description of quantization and interpolation we ignored pitch-doubling and pitch-halving phenomena. Such situations require special treatment [3,4]; for interpolation, the shorter prototype waveform is repeated; for differential quantization, the previous quantized waveform is either repeated or only half of the waveform is used.

For voiced speech segments with a high level of aspiration noise, synthesis with (10) sometimes results in buzziness, because of too high a level of periodicity in higher frequency ranges. Adding a noise signal with amplitude modulation based on the power envelope of $x(t)$ [10] removes these artifacts. The noise energy should be frequency-dependent, and the energy of the prototype waveforms should be reduced to account for this frequency-dependent noise energy. For best results the noise power should be derived from robust measurements of the correlations between adjacent pitch cycles in various frequency bands. However, surprisingly good results can be obtained by adding a noise signal of fixed statistics, modulated according to the signal power of $x(t)$, to all voiced speech segments.

RESULTS

The PWI coding algorithm is illustrated in Figure 1. Figure 1(a) shows the original speech waveform for a voiced interval and 1(b) the pitch markers. The PWI coding procedure reduces to LP vocoding if only single impulses are used for the prototype excitation waveforms. Figures 1(c) and 1(d) show the resulting excitation and reconstructed speech waveforms. Note that, in contrast with conventional analysis-by-synthesis coders such as CELP, the signal reconstructed with the PWI coder is not synchronous with the original speech signal. The time location of the pitch pulses is a function of the initial conditions in the first frame coded with the PWI coder and the pitch contour.

In the case of Figures 1(c) and 1(d), each prototype waveform is represented by its pitch period, its impulse amplitude, and a set of LP coefficients. By using 7, 8, and 24 bits [11], respectively, for these parameters, in combination with a 25 ms update interval, an overall bit rate of 1.6 kb/s is obtained. As expected, at this bit rate PWI achieves only a vocoder-like speech quality and suffers from some buzziness.

Better quality is obtained when the waveform shape is described too. Figures 1(e) and 1(f) show the excitation and reconstructed speech waveforms, at an overall bit rate of 2.5 kb/s. In this case the prototype waveforms are differentially encoded, using two codebooks of 8 bits each. 15 bits are used for the gains of the codebooks and the previous prototype waveform. In the reconstructed speech of the 1.6 kb/s example the buzziness is almost completely removed. Introduction of the modulated noise signal with adapted statistics completely removes any remaining buzziness. For comparison, Figures 1(g) and 1(h) show the excitation and the reconstructed speech waveforms obtained for the unquantized case.

Since the PWI coder is used for voiced sections only, with another coder being used for the unvoiced signals, the first interpolation interval lacks a past prototype waveform. This past prototype waveform can be obtained by either replicating the first transmitted prototype waveform, or by extraction of a prototype waveform from the previous frame of reconstructed speech. In the former case, proper alignment at the transition is determined by cross correlation of the reconstructed signals. In our implementations, we extracted a prototype waveform from the previous frame of reconstructed speech. A proper voiced-to-unvoiced transition is straightforward. The original signal must be displaced such that the end of the last prototype waveform corresponds to the beginning of the first unvoiced frame.

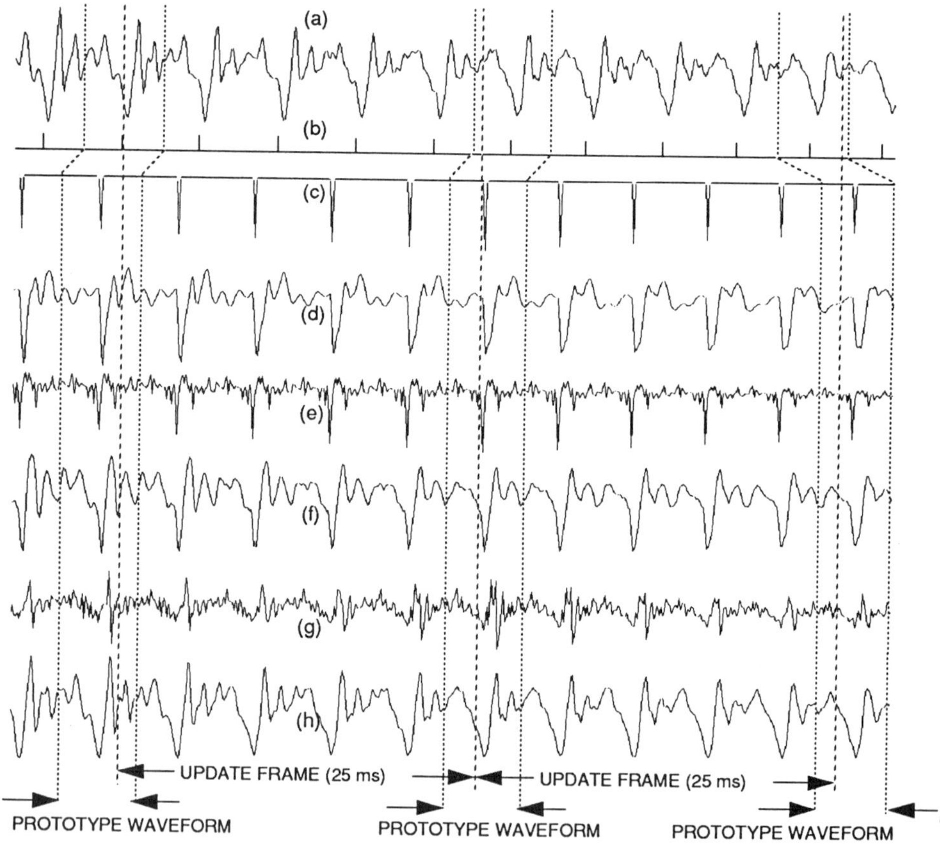

Figure 1. PWI encoding of speech. (a) original signal, (b) pitch-pulse markers, (c) and (d) 1.6 kb/s PWI excitation and associated speech signal, (e) and (f) 2.5 kb/s PWI excitation and associated speech signal, (g) and (h) unquantized PWI excitation and associated speech signal.

At a first frame to be coded with the PWI coder, the past prototype waveform for interpolation is to be distinguished from the past prototype waveform used for differential quantization. It is advantageous to define a single pulse as the past prototype waveform for quantization in the first frame to be coded with the PWI coder. This way differential encoding can be used even for the first prototype waveform to be transmitted.

Mean-Opinion Score (MOS) listening tests were performed in which the voiced segments of the speech reconstructed by several coders was replaced with a 2.5 kb/s PWI-coded signal. When the voiced segments were replaced by a PWI-coded signal in speech coded by the new 16 kb/s CCITT standard [12], no statistically significant effect on the MOS score was found. When the same voiced segments were replaced by a PWI-coded signal in speech coded with a 4.8 kb/s CELP algorithm, a significant increase in the MOS score was obtained. In the latter combination of coders, the transitions from unvoiced to voiced, where the waveform is determined by the CELP algorithm, were the source of most audible distortion.

CONCLUSIONS

The prototype-waveform interpolation procedure provides a more efficient method for coding voiced speech than conventional analysis-by-synthesis procedures. The main reason for this efficiency is a relaxation of certain waveform-matching constraints which are implicit in these conventional methods, but which are perceptually not significant. In particular, in the PWI algorithm the accuracy of waveform matching is independent of the pitch contour and of the level of periodicity.

The periodicity is critical to the perceived quality of voiced speech. Generally the correlation between adjacent pitch cycles is high at low frequencies and decreases at higher frequencies. When the bit rate of conventional analysis-by-synthesis based coders is lowered, the periodicity and, therefore, the speech quality decreases. In the PWI algorithm the accuracy of matching of the prototype waveform shape decreases with decreasing bit rate. However, because the excitation signal is reconstructed by means of interpolation, the periodicity of the signal does not go down when the matching accuracy decreases. As a result, the perceived quality degrades more gracefully with decreasing bit rate.

Since the signal is reconstructed from a downsampled sequence of pitch cycles (one prototype waveform every 20-30 ms), no waveform matching is performed in the regions between the extracted prototype waveforms. While this is advantageous for most voiced speech signals, the periodicity assumption means a lowered robustness against nonspeech sounds. The concept of generalized analysis-by-synthesis [3, 13] offers recourse against such problems. In this paradigm, the original signal is modified so as to maximize coder performance. The modifications are constrained to be perceptually insignificant. Here, the original signal would be time warped to match the PWI-reconstructed signal. The PWI-reconstructed signal can then be corrected with conventional CELP techniques to obtain a better match to the modified original signal.

References

[1] P. Kroon and Ed. F. Deprettere, ''A Class of Analysis-by-Synthesis Predictive Coders for High Quality Speech Coding at Rates between 4.8 and 16 kbit/s.,'' *IEEE J. Selected Areas Comm.* **6** pp. 353-363 (1988).

[2] B. S. Atal and M. R. Schroeder, ''Predictive Coding of Speech Signals and Subjective Error Criteria,'' *IEEE Trans. Speech Signal Proc.* **ASSP-27**(3) pp. 247-254 (1979).

[3] W. B. Kleijn, *Analysis-by-Synthesis Speech Coding Based on Relaxed Waveform-Matching Constraints,* Ph.D. thesis, Delft University of Technology, Delft, The Netherlands (1991).

[4] W. B. Kleijn and W. Granzow, ''Methods for Waveform Interpolation in Speech Coding,'' *Digital Signal Processing* **1**(4) pp. 215-230 (1991).

[5] W. B. Kleijn, ''Continuous Representations in Linear Predictive Coding,'' *Proc. Int. Conf. Acoust. Speech Sign. Process., Toronto,* pp. 201-204 (1991).

[6] W. Verhelst, ''On the Quality of Speech Produced by Impulse Driven Linear Systems,'' *Proc. Int. Conf. Acoust. Speech Sign. Process., Toronto,* pp. 501-504 (1991).

[7] W. Granzow, B. S. Atal, K. K. Paliwal, and J. Schroeter, ''Speech Coding at 4 kb/s and Lower Using Single-Pulse and Stochastic Models of LPC Excitation,'' *Proc. Int. Conf. Acoust. Speech Sign. Process., Toronto,* pp. 217-220 (1991).

[8] J. Haagen, H. Nielsen, and S. Hansen, ''Improvements in 2.4 kbps High-Quality Speech Coding,'' *Proc. Int. Conf. Acoust. Speech Sign. Process., San Francisco,* pp. II145-II148 (1992).

[9] W. Hess, *Pitch Determination of Speech Signals,* Springer Verlag, Berlin (1983).

[10] D. J. Hermes, ''Synthesis of Breathy Vowels: Some Research Methods,'' *Speech Communication* **10** pp. 497-502 (1991).

[11] K. K. Paliwal and B. S. Atal, ''Efficient Vector Quantization of LPC Parameters at 24 Bits/Frame,'' *Proc. Int. Conf. Acoust. Speech Sign. Proc. Toronto,* pp. 661-664 (1991).

[12] J-H. Chen, ''A Robust Low-Delay CELP Speech Coder at 16 kb/s,'' pp. 25-35 in *Advances in Speech Coding*, ed. B. S. Atal, V. Cuperman, A. Gersho, Kluwer Academic Publishers, Dordrecht, Holland (1991).

[13] W. B. Kleijn, R. P. Ramachandran, and P. Kroon, ''Generalized Analysis-by-Synthesis Coding and its Application to Pitch Prediction,'' *Proc. Int. Conf. Acoust. Speech Sign. Process., San Francisco,* pp. I337-I340 (1992).

PART V

AUDIO CODING

Emerging applications, such as audio and video teleconferencing and high fidelity audio transmission and entertainment products, have motivated considerable interest in new audio coding problems that depart from the traditional single channel telephone bandwidth signal coding task. In particular, there is an expanding interest in digital coding of wide-band speech and audio signals for bandwidths ranging from 7 kHz up to 20 kHz. Most of the contemporary techniques for coding telephone bandwidth speech are applicable with suitable modifications to 7 kHz signals. However, substantially distinct coding techniques are generally needed for high fidelity audio signals with bandwidths of 15 or 20 kHz. In this section, a representation of different audio coding techniques is presented. Harborg et al. describe a CELP-based 16 kb/s coder for 7 kHz audio suitable for videophone applications. Champion et al. consider the application of multi-rate sinusoidal transform coding (STC) to an audio conferencing bridge. Shoham reports on a 7 kHz audio coder at 32 kb/s based on a delayed-decision CELP technique. De Iacovo et al. present a split band CELP coder for 7 kHz coding at 16 kb/s. LaFlamme et al. describe a 7 kHz audio coder at 9.6 kb/s based on CELP coding with algebraic excitation codebooks. Finally, the coding of high fidelity 15 kHz audio signals based on transform coding with generalized product code (GPC) vector quantization methods is presented by Chan and Gersho.

15

A WIDEBAND CELP CODER AT 16 kbit/s FOR REAL TIME APPLICATIONS

Erik Harborg and Arild Fuldseth
SINTEF DELAB, N-7034 Trondheim, NORWAY

Finn Tore Johansen and Jan Eikeset Knudsen
Norwegian Telecom Research, P.O.Box 83, N-2007 Kjeller, NORWAY

INTRODUCTION

Since its introduction in 1984, Code Excited Linear Predictive (CELP) [1] coding has received considerable attention for high quality speech coding at low bit-rates. Although most of the research has been focused on coding of narrowband (200-3400 kHz) speech, some recent studies on CELP coding of wideband (50-7000 kHz) speech have been reported [2], [3], [4].

A possible application for wideband speech coders is the loudspeaking videophone, where it is foreseen that for the next generation videophones, a 64 kbit/s channel can be used for both speech and video. Here, we present results on a 16 kbit/s, 7 kHz CELP coder which will allow wideband speech within a 64 kbit/s videophone service.

BASIC CODER STRUCTURE

In CELP coding the synthesized speech signal is constructed by feeding an excitation signal through an LPC synthesis filter $1/A(z)$. For the coder presented here, the excitation signal is constructed by selecting one or more adjacent vectors from an adaptive codebook where the codebook vectors overlap in all samples but one, and one vector from a fixed stochastic codebook with non-overlapping vectors. The selected vectors are scaled by their respective gain factors and added together to form the final excitation signal. The adaptive codebook contains the past history of the excitation signal itself and is an implementation of a pitch synthesis filter where the codebook index corresponds to the pitch period. Note also that if more than one vector is selected from the adaptive codebook, this is an implementation of a higher-order pitch synthesis filter. The stochastic codebook contains sparse codevectors with three non-zero pulses generated by a Gaussian process.

The CELP encoder which is illustrated in Figure 1, processes the input speech signal in three steps; the LPC-analysis, the adaptive codebook search, and the stochastic codebook search. All codebook vectors are selected by an analysis-by-synthesis procedure where the weighted mean squared error between the input signal and the synthesized speech signal is minimized using a noise weighting filter $A(z)/A(z/\gamma)$.

The weighting factor γ is set to 0.9 for the stochastic codebook search and to 0.6 for the adaptive codebook search. Further, we use the autocorrelation and the covariance error criterion [5] for the adaptive and the stochastic codebook search

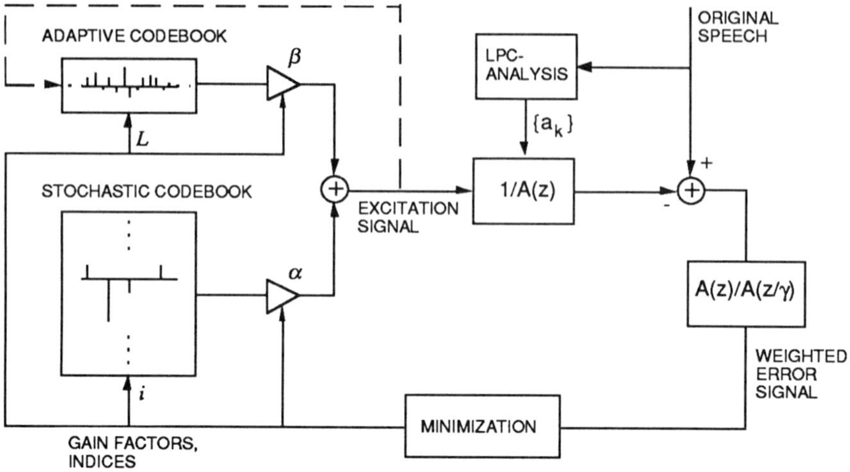

Figure 1 CELP encoder.

respectively. The LP-coefficients are found by Burg's algorithm, and are represented by Log-Area-Ratios (LAR) for quantization purposes.

ADAPTIVE AND STOCHASTIC CODEBOOK SEARCH

With a $(2K+1)$th order pitch synthesis filter implemented as an adaptive codebook, the optimum pitch period L (or codebook index), and the pitch coefficients, β_i should be chosen to minimize the following error criterion

$$\varepsilon = \left(s_1^T - \sum_{i=-K}^{K} \beta_i e_{L+i}^T H^T\right)\left(s_1 - \sum_{i=-K}^{K} \beta_i H e_{L+i}\right) \tag{1}$$

where s_1^T is the target vector, e_j^T are the codebook vectors and H is the impulse response matrix of the weighted synthesis filter, 1/A(z/γ). The target vector s_1, is formed by filtering the input speech through the weighting filter A(z)/A(z/γ), and then subtracting the zero input response of 1/A(z/γ).

The optimum set of pitch parameters can be found by optimizing the pitch period and the pitch coefficients simultaneously with the coefficient quantization procedure within the search loop. However, in order to reduce the computational load, we use a suboptimum two-step search procedure, where the pitch period is found in the first step, using a conventional first order search (K=0) without coefficient quantization. Thus, the pitch period, L is the value of j which maximizes

$$(s_1^T H e_j)^2 / (e_j^T H^T H e_j) \qquad L_{min} \le j \le L_{max} \tag{2}$$

In the second step, the pitch coefficient vector is determined by use of vector quantization. Here the optimum vector is selected from a pitch coefficient codebook as the vector minimizing the error criterion in Eq. (1) given the pitch period L.

The pitch coefficient codebook is trained by applying the LBG algorithm [7] with the distortion measure in Eq. (1) for classification of the training data [8].

The coder performance for various orders of the pitch synthesis filter and various numbers of quantizer bits was evaluated using 180 seconds of speech from four speakers (two male and two female). The resulting segmental SNR values are listed in Table 1. For these simulations the coder configuration in Table 2 with 90 bits for the LP-coefficients was used.

# bits	SNRSEG [dB]		
	1st order	3rd order	5th order
unquantized	14.26	15.19	15.92
9	14.23	15.08	15.42
7	14.29	14.91	15.15
5	14.18	14.63	14.74
3	13.88	13.96	14.04
1	12.59	12.62	12.71

Table 1 Segmental SNR values for various orders of the pitch synthesis filter and numbers of quantizer bits.

For the stochastic codebook search, the distortion measure can be expressed as

$$\varepsilon = \|s_2 - \alpha_i H c_i\|^2 = (s_2^T - \alpha_i c_i^T H^T)\,(s_2 - \alpha_i H c_i) \tag{3}$$

where α_i is the optimum gain factor. Here, the target vector s_2 is obtained by subtracting the contribution from the adaptive codebook from s_1. The optimum codebook vector c_i is then determined as the vector which maximizes

$$(s_2^T H c_i)^2 / (c_i^T H^T H c_i) \tag{4}$$

The gain factor α_i, is encoded by using linear prediction from frame to frame in the log domain, and is selected such as to minimize the error in Eq. (3) given the codebook vector c_i.

Parameters	Frame	Update rate [Hz]	#bits/frame	Bit-rate [b/s]
LPC (20 coeff.)	20 ms	50	70	3500
pitch coef. (β)	2 ms	500	5	2500
pitch period (L)	2 ms	500	8	4000
stoch. gain (α)	2 ms	500	5	2500
stoch. index (i)	2 ms	500	7	3500
Total				16000

Table 2 Coder configuration at 16 kbit/s (5th order pitch synthesis filter, 16 kHz sampling).

CODER CONFIGURATION AND COMPLEXITY

The configuration for a version of the CELP coder using a 5th order pitch synthesis filter is shown in Table 2. The computational effort of the CELP coder is dominated by the computation of the numerator and the denominator terms of Eqs. (2) and (4) for every vector in the codebooks. A direct evaluation of Eqs. (2) and (4) is computationally very expensive. However, by utilizing the efficient search methods as described in [3] [5] [6], the number of instructions (multiply-accumulate, multiply, add) per second can be reduced considerably. These methods utilize the overlapping structure of the adaptive codebook, the sparsity of the stochastic codebook vectors, and the fact that the codebooks are searched several times between each update of the LP-coefficients. For the coder configuration in Table 2 the computational load for the encoder/decoder is estimated at 14.9 million instructions per second (MIPS) as indicated in Table 3.

Task	MIPS
LPC analysis	2.42
adaptive codebook search	6.49
stochastic codebook search	0.48
other	5.51
Total	14.90

Table 3 Computational load for encoder/decoder.

LISTENING TEST

In order to perceptually evaluate the proposed CELP coder a listening test has been performed comparing the CELP coder with the CCITT G.722 sub-band coder at 48, 56 and 64 kbit/s. The following coders participated in the test:

CELP: CELP, configured as in Table 2, 16 kbit/s
G722_48: CCITT standard G.722 sub-band coder at 48 kbit/s
G722_56: CCITT standard G.722 sub-band coder at 56 kbit/s
G722_64: CCITT standard G.722 sub-band coder at 64 kbit/s

The test procedure followed the Absolute Category Rating Method as described in [9]. The test was conducted in Norwegian, and a scale from 1 to 5 was used. There were 8 talkers (4 male + 4 female) and 16 listeners. In evaluating wideband speech the choice of listening device is of great importance and may influence the ranking of coders. While high quality headsets was used for optimization and selection of parameters, the ACR test was performed using loudspeaker listening which will probably be more realistic in a real videotelephony situation. The results are given in Figure 2. In this test a significant difference is less than 0.2 on the MOS scale at 95% confidence level.

CONCLUSION

We have presented a low-complexity wideband CELP coder running at 16 kbit/s which can be implemented in real-time on a single DSP. The performance of the coder

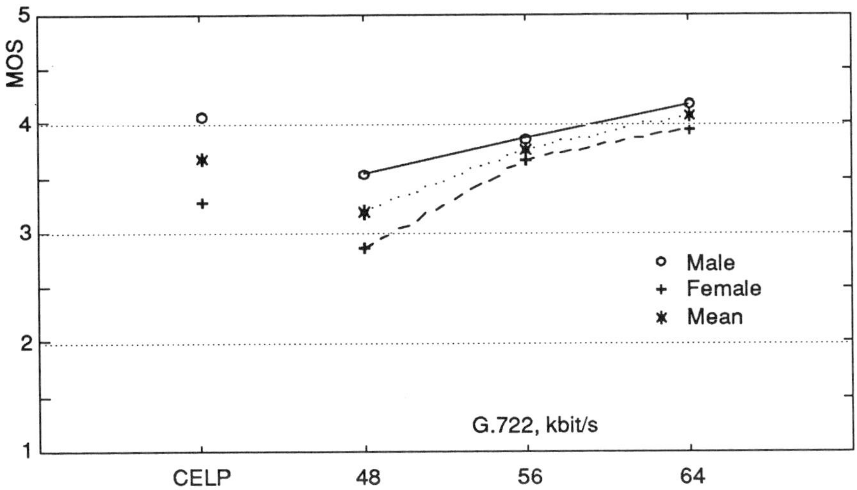

Figure 2 MOS scores for female and male.

has been compared to the CCITT G.722 sub-band coders at 48, 56 and 64 kbit/s using an Absolute Category Rating test. It was found that the CELP coder is comparable to the G.722 coder at 56 kbit/s, however with a larger difference between male and female speakers at this bit rate.

REFERENCES

[1] B.S. Atal, M.R. Schroeder: "Stochastic coding of speech signals at very low bit rates," Proc. IEEE Int. Conf. Communications 1984.

[2] R. Drogo de Jacovo, R. Montagna, F. Perosino, D. Sereno: "Some experiments of 7 kHz audio coding at 16 kbit/s," Proc. ICASSP 1989.

[3] A. Fuldseth, E. Harborg, F.T. Johansen, J.E. Knudsen: "A real-time implementable 7 kHz speech coder at 16 kbit/s," Proc. EUROSPEECH 91.

[4] C. Laflamme, J.-P. Adoul, R. Salami, S. Morisette, P. Mabilleau: "16 kbps wideband speech coding technique based on algebraic CELP," Proc. ICASSP 1991.

[5] W.B. Kleijn, D.J. Krasinski, R.H. Ketchum: "Fast methods for the CELP speech coding algorithm," IEEE Trans. on ASSP, vol. 38, no. 8, Aug. 1990.

[6] C. Laflamme, J.-P. Adoul, H.Y. Su, S. Morisette, "On reducing computational complexity of codebook search in CELP coder through the use of algebraic codes," Proc. ICASSP 1990.

[7] Y. Linde, A. Buzo, R.M. Gray: "An algorithm for vector quantizer design," IEEE Trans. on Comm., vol. COM-28, no. 1, 1980.

[8] A. Fuldseth, E. Harborg, F.T.Johansen, J.E.Knudsen: "Pitch prediction in a wideband CELP coder," Proc. EUSIPCO 1992.

[9] CCITT Draft Recommendation P.80, COM XII-52E, part B, July 1990.

16

MULTIRATE STC AND ITS APPLICATION TO MULTI-SPEAKER CONFERENCING[1]

Terrence G. Champion
COMSEC Engineering Office
RL/ERT, Hanscom AFB, MA 01731-5000

Robert J. McAulay and Thomas F. Quatieri
Lincoln Laboratory, MIT
244 Wood Street
Lexington, MA 02173-9108

INTRODUCTION

The problem of conferencing over systems which employ parametric vocoders has long been of interest to the military. In analog or wideband digital conferencing, overlapping speakers are handled by signal summation at a conferencing bridge. Such a scheme is not feasible for parametric vocoders which would require synthesis and reanalysis of the aggregate speech signal, a process called tandeming, which results in severe loss in quality in the synthetic speech. Moreover, further degradations occur when multiple speakers are active since parametric vocoders are not designed to model more than one voice. One narrowband technique currently in use is based on the idea of signal selection—a speaker has the channel until finished or until replaced by someone with a higher priority, and speakers contend for the open channel when it becomes available [1]. The advantage of such a technique is that it avoids the degradations due to tandeming, but it is cumbersome. A more natural conference control is handled by interruptions corresponding to multiple speakers producing overlapping speech. One scheme that permits two-speaker overlaps assigns one-half of the available bandwidth to each speech coder and defers signal summation to the terminal [2]. This approach limits the overall quality of the conference by forcing the coder to work at half the bandwidth. Since for the majority of a conference there will be only a single active speaker, this technique causes an overall degradation in the perceived quality in order to model an event that occurs relatively infrequently.

The technique proposed here also defers signal summation to the terminal, however, it adaptively allocates the available bandwidth based on the number of active speakers. Since during most of a conference there will only be a single speaker, the quality of the speech will be maintained at the highest level and this maintains the perceived quality of the conferencing system. When there are two speakers present, the speech quality of the individual speakers will be somewhat reduced; however, since each speaker is allocated one-half of the

[1]This work was sponsored by the Dept. of the Air Force. The views expressed are those of the authors and do not reflect the official policy or position of the U.S. Government.

bandwidth, the intelligibility of the two speakers can be preserved, and if no significant artifacts occur when there are two speakers, then the method should allow for a more natural contention for the conference control.

MULTIRATE SINUSOIDAL TRANSFORM CODER

It has been shown that speech of very high quality can be synthesized using a sinusoidal model when the amplitudes, frequencies and phases are derived from a high-resolution analysis of the short-time Fourier transform (STFT), [3]. It has also been shown that if the measured sine-wave frequencies are replaced by a harmonic set of frequencies in which the fundamental frequency is chosen to make the harmonic model a "best fit" to the measured sine-wave data, then synthetic speech of high quality can also be obtained provided the amplitudes and phases are obtained by sampling the STFT at the harmonic frequencies [4]. A model has also been developed for the sine-wave phases which has a linear component corresponding to the onset time of the glottal pulse, a minimum phase component due to the dispersive characteristics of the vocal tract and a random component that represents the degree to which the speech segment is unvoiced [5]. The parameters of the resulting speech model are the pitch, voicing and the sine-wave amplitudes at the pitch harmonics.

While conventional methods would be used for coding the pitch and voicing, new methods have been developed for coding the sine-wave amplitudes [6]. Basically, these are based on fitting a set of cepstral coefficients to an envelope of the measured sine-wave amplitudes. The advantage of the cepstral coefficients over all-pole modelling, for example, is the fact that they assume no constraining model shape, except that the vocal tract be minimum phase. This results in better fits to the amplitudes in the baseband region which seems to be important in retaining speaker naturalness. Moreover, the cepstral model adds the desired dispersive characteristics to the sine-wave phases. This is particularly important during a mixed voiced-unvoiced speech segment, since the randomness of the sine-wave amplitudes is reflected into the system phase through the minimum phase assumption. The added randomness in the phases of the sine waves contributes to naturalness in the synthetic speech.

A variety of methods have been studied for coding the cepstral coefficients [6], and this continues to be an interesting area of research [7]. The resulting system, referred to as the Sinusoidal Transform Coder (STC), was found to produce natural sounding speech with more or less uniformly increasing quality from 2400 b/s to 4800 b/s. In fact, in a recent TIA vocoder pre-selection test, STC at 4.0 kb/s was shown to yield an MOS score statistically equivalent to that of the full-rate VSELP codec running at 8.0 kb/s [8]. An informal test of STC at 2.4 kb/s has shown that performance is about one-half MOS point less than that of the 4.0 kb/s STC system. Since STC is a parametric vocoder that depends on pitch, voicing, and spectral envelope information, it not only lends itself to quantization at low data rates, but it is amenable to transformation of the parameters from one rate to another with relative ease. In fact, a simulation has been developed that allows for conversion from 4800 b/s to 2400 b/s, without the introduction of artifacts at the transition frames. This initial demonstration was done simply by requantizing the cepstral coefficients at the lower rate, applying frame-fill to the pitch and holding all other parameters constant. It is this multirate capability that is important to the conferencing application.

MULTI-SPEAKER CONFERENCING

The conferencing system under development at Rome Laboratory consists of two devices: a speech terminal for each conferee and a conferencing bridge. The speech terminal performs the vocoder analysis and synthesis functions. The terminal always performs speech analysis at the highest rate allowed by the channel. During analysis the speech terminal also makes a determination as to whether or not the conferee is actually speaking, and codes a voice-activity bit into the data stream.

Since in this conferencing system signal summation is being deferred to the speech terminal, synthesis is complicated by the fact that there will often be multiple signals to be synthesized and summed. When there are multiple speakers, the synthesizer must have the capability to represent them. One approach would be to synthesize each speaker separately and to sum the time-domain signals. The use of STC makes possible a much simpler technique since the summation can be done in the parameter domain prior to synthesis, greatly reducing the computational complexity of the synthesizer. This is accomplished using the overlap-add sine-wave synthesis technique [9]. With this method the pitch, voicing and spectrum are used to compute the amplitudes, frequencies and phases of the underlying sine waves, and these are used to fill in the complex FFT buffer at the sine-wave frequencies. The speech waveform is obtained from the inverse transform with the overlap-add procedure being used to smooth transitions across frame boundaries. With two speakers, the two sets of complex parameters are added in the transform domain before taking the inverse transform. In this way synthesis of two speakers involves only slightly more computation than for one speaker.

The bridge has two basic functions: (1) signal routing; and (2) bit-rate reduction on speaker parameter sets to allow for multiple speakers to be transmitted through the channel. When there is only one active speaker, all conferees (except the active speaker) receive the same set of parameters. When there are two active speakers, each speaker would receive the other speaker's parameters at the highest rate, while the passive listeners would receive the two parameter sets of the two active speakers, each transformed to a lower bit-rate. Figure 1 shows a typical scenario with three conferees, two of which are actively speaking. The idea of splitting the channel to allow for the parameter sets of multiple speakers depends on an effective transformation from a higher bit-rate to a lower bit-rate. The dynamic multirate capability of STC lends itself naturally to the implementation of this transformation process. In addition, STC seems to be far less sensitive to frame rate than other narrowband vocoder algorithms, a property that allows the designer a great deal of freedom when designing interoperable systems working at different data rates.

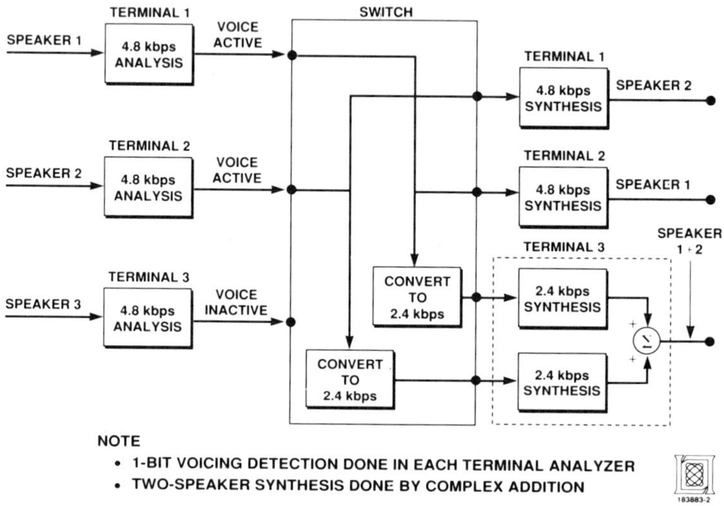

Figure 1: **Three conferees with two interrupting speakers.**

CONFERENCING BRIDGE SIMULATION

As of this writing, the conferencing bridge has been implemented in non-real-time. This version of the bridge can handle up to four conferees with two speakers active at any one time. The highest rate used is 4.8 kb/s for a single speaker, 2.4 kb/s for two speakers.

The control logic for the bridge is fairly simple. Two slots are available for active speakers on a first-come, first-serve basis. New speakers that begin while both slots are occupied are denied access to a newly-freed channel to prevent active speakers from being interrupted in mid-sentence. Since some interpolation of parameters is done, care must be taken to properly associate parameters going into and out of collisions. For this purpose the bridge recognizes and codes one of four states. One state represents no change from the previous state. Another state signals an increase in the number of speakers from one to two (one speaker is assumed); the other two states identify which speaker is still speaking during the translation back to one speaker.

A real-time implementation of this technique is currently under development to evaluate the overall acceptability of the method and to optimize bridge control structures. Depending upon the outcome of these experiments, further research may be devoted to the development of a system allowing for more conferees and for a three-speaker collision.

References

[1] J.W. Forgie, C.E. Feehrer, and P.L. Weene, "Voice Conferencing Technology Problem," MIT Lincoln Laboratory Final Report, 31 March 1979.

[2] D. Busson, N. Irisarry, and C. Stengel, "Secure Conferencing HF Communications," RADC-TR-86-55, April 1986.

[3] R.J. McAulay and T.F. Quatieri, "Speech Analysis/Synthesis Based on a Sinusoidal Representation," *IEEE Trans. ASSP*, Vol. ASSP-34, No. 4, pp. 744–754, 1986.

[4] R.J. McAulay, and T.F. Quatieri, "Pitch Estimation and Voicing Detection Based on a Sinusoidal Model," *IEEE Proc. Int. Conf. Acoustics, Speech and Signal Processing 1990*, Albuquerque, NM, pp. 249–252, April 1990.

[5] R.J. McAulay, and T.F. Quatieri, "Sine-Wave Phase Modelling at Low Data Rates," *IEEE Proc. Int. Conf. Acoustics. Speech and Signal Processing 1991*, Toronto, Canada, May 1991.

[6] R.J. McAulay and T.F. Quatieri, "Low-Rate Speech Coding Based on the Sinusoidal Model," Chapter 1.6, pp. 165–207, in *Advances in Acoustics and Speech Processing*, M. Sondhi and S. Furui, Eds., Marcel Deckker, 1992.

[7] R.J. McAulay and T.F. Quatieri, "The Sinusoidal Transform Coder at 2400 b/s," to be published in *Proc. MILCOM'92*, San Diego, CA, October 1992.

[8] D. Lin, "Statistical Analysis of the BNR Half-Rate MOS Data Set," TIA Speech Codec Working Group, Toronto, June 1992.

[9] R.J. McAulay and T.F. Quatieri, "Computationally Efficient Sine-Wave Synthesis and Its Application to Sinusoidal Transform Coding," *IEEE Proc. Int. Conf. Acoustics, Speech and Signal Processing 1988*, New York City, NY, April 1988.

17

LOW DELAY CODING OF WIDEBAND SPEECH AT 32 KBPS USING TREE STRUCTURES

Yair Shoham

Speech Coding Research Department
AT&T Bell Laboratories
600 Mountain Ave.
Murray Hill, NJ 07974

INTRODUCTION

The prospect of high-quality commentary-grade multi-channel/multi-user speech communication via the emerging ISDN has raised a lot of interest in advanced coding algorithms for 50-7000 Hz wideband speech. A high-quality 32Kbps wideband speech coder has recently been developed in our laboratory [1,2]. This coder is based on the Low-Delay Code-Excited Linear-Predictive (LD-CELP) algorithm. It employs 5-sample vector quantization (VQ) with an end-to-end delay of only about 0.94 msec. Its performance, as judged by informal listening tests, is comparable to that of the 64Kbps standard (G.722) CCITT wideband coder [3]. Since a much longer delay can be tolerated in many (if not all) wideband-speech applications [4], it is possible, in principle to further improve the performance by increasing the frame size and the coding delay. A straightforward extension of the frame size, however, implies an exponential increase of coding complexity that is characteristic of VQ-based algorithms.

In the study reported here, we have investigated the incorporation of the LD-CELP algorithm in a delayed-decision coding (DDC) framework as one possible method for increasing the delay with a *linear* rather than exponential growth of complexity. The proposed coder combines short-frame VQ with long-frame tree structures, based on the ML-algorithm [5]. Hence, it will be referred to as low-delay vector-tree CELP (LDVT-CELP) coder.

This work shows that LDVT-CELP outperforms the basic LD-CELP coder at the price of a longer delay and a linear increase of complexity.

THE BASIC WIDEBAND-SPEECH LD-CELP CODER

The basic wideband LD-CELP coder is shown in Figure 1. At the transmitter, a codebook of excitation vectors is used to excite a cascade of an LPC-derived all-pole filter $1/A(z)$ and a noise-shaping filter $W(z)$. The output Y of the combined filter is compared to a filtered version X of the input speech S. X is obtained from S by the same filter $W(z)$. The notations $\hat{S}, S, X, Y$ all stand for K-sample vectors. The difference norm $\| X - Y \|$ is computed for all possible scaled excitation vectors $g\, C_j,\ j=1,..,N$ where N is the codebook size. The index j corresponding to the norm-minimizing excitation vector is sent over to the receiver. The receiver retrieves the jth excitation vector and duplicates the synthesis of the coded speech $\hat{S}$.

The LPC filter is derived from the *quantized* speech. This backward LPC analysis is fundamental to this low-delay coder. The excitation scale factor (gain) is also computed in a backward mode, namely, predicted from past quantized data. The noise-shaping filter, which is a critical component in this coder, is of the form:

$$W(z) = \frac{B(z/\gamma_z)}{B(z/\gamma_p)\, T(z/\gamma_t)} \tag{1}$$

$B(z)$ is the standard LPC polynomial obtained from the *input* signal. Since the frame is short, it is advantageous to perform these analyses in a recursive mode [8]. $T(z)$ is a low-order polynomial that captures the tilt of the smoothed LPC spectrum. It is derived from $B(z)$ by applying the standard LPC analysis to the unit-sample response of $B(z)$. The pole-zero section $B(z/\gamma_z) \,/\, B(z/\gamma_p)$ de-emphasizes the formants and emphasizes the inter-formant regions by a proper selection of γ_z and γ_p. The section $1/T(z)$ emphasizes high frequencies to a degree controlled by γ_t. The shaping parameters used are γ_z=0.98 γ_p=0.8 γ_t=0.7. The orders of $A(z)$ and $B(z)$ are 32 and 16, respectively. A 10-bit codebook is used in this coder, with a frame size of 5 samples. This corresponds to the bit rate of 32 Kbps at a sampling rate of 16000 Hz.

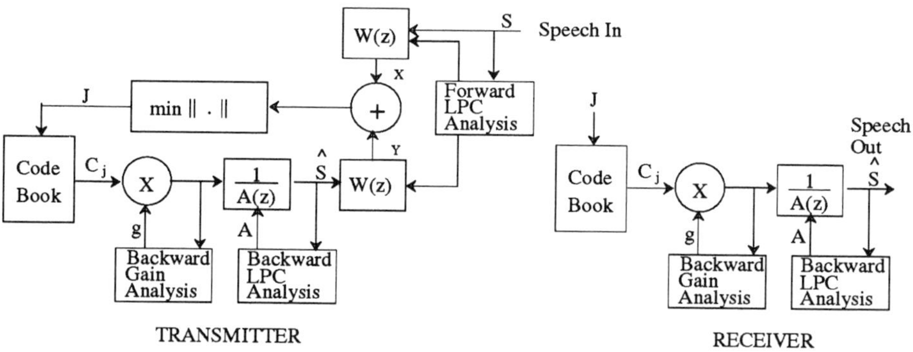

Figure 1. Basic Low-Delay CELP

This coder delivers high performance [1,2], equivalent to that of the 64Kbps CCITT standard coder (G722). The objective was to push this performance closer to transparent quality by combining the coder with tree coding, as described below.

LINEAR-PREDICTIVE VECTOR TREE CODING

The LD-CELP was incorporated into a directed tree structure governed by the ML-algorithm. A directed tree is built of nodes and branches that leave one node and arrive in another. A path in a directed tree is a connected sub-tree for which each node has no more than one arriving branch and no more than one leaving branch. In this work, we deal with *real* trees for which each node has one and only one arriving branch, which implies that each node is an end-point of one and only one path. A node is usually associated with a *state* of the system as evolved by traveling via that particular path. A branch is associated with an action to be taken to move the system from the current state to the new one, and with the cost incurred by doing so.

In the context of vector-predictive speech coding, a state (node) of the system is the ensemble of all the variables that determine the dynamic behavior of the coder, one of which is the coder output, namely, the quantized speech. Also, associated with a node is the accumulated distortion of the speech waveform incurred by traveling through the corresponding path. The usual mean-square error (MSE) was used here to measure this distortion. A branch is associated with a codevector (or a codeword index) to be used in synthesizing a new state for one frame speech, given the old one, and with the incremental distortion resulted from doing so. Each node is also associated with a time index. The tree is constrained to be causal and incremental, that is, a branch leaving a node with time-index l can only arrive in a node having time index $l+1$. The time index is also referred to as a *depth* into the tree.

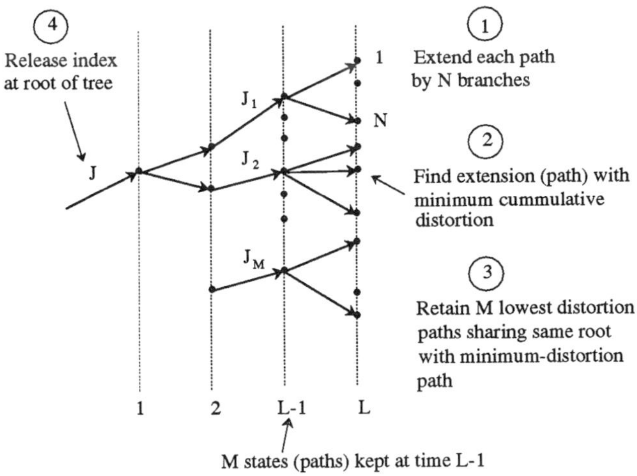

Figure 2. Basic ML tree pruning Algorithm

The ML-algorithm, shown in figure 2, maintains only up to M paths at any depth. The rest are discarded, pruning the otherwise-ever-growing tree to a fixed width of M nodes. Moreover, the M survivor paths are so selected as to branch off from the same node (path) L time-indices back. This ambiguity-removing constraint is essential in source coding and transmission since the receiver, while duplicating the transmitter, can follow only one path. Given the M paths from time $l=L-1$, the coder first extends each path by N branches corresponding to all entries in a given excitation codebook. An array of MN distortions is produced, one for each of $M\,N$ extension candidates. It is intuitively clear that the extended path with the minimum accumulated cost should be retained. A central issue in this type of tree coding is what other $M-1$ paths should be kept. In this work, we employed the standard strategy of keeping the M out of $M\,L$ lowest-accumulated-distortion paths, subject to the ambiguity test mentioned above where the root common to all survivor paths is the node on the *best* path at time $l=1$. Once the pruning is done at time $l=L$, the path ending at the common node at time $l=1$ is no longer subject to a possible deletion. The codeword indices associated with its branches can be transmitted. This is actually done by releasing one root-index at a time - a mode called *incremental release*.

There are other modes in which more than one index are released at one time. One particular is the *block release* mode in which L indices of times $l=1,..,L$ are released as a block. These indices correspond to the best path ending at a node at time $l=L$. At this time, the tree collapses to a width of one, leaving only one (best) path. It take another L steps to build a new tree and to release the next set of L indices. We experimented with both incremental and block release modes. Interestingly, the incremental mode was always the better one. This is explained by the blockiness effect in the block mode, namely, the dependencies between paths inside a block and those of the next block are neglected, which degrades the performance.

Running the LD-CELP, described in the previous section, on a tree means maintaining the states of M different coders since each node represents a state of a coder with a different history. The structure of the LD-CELP implies keeping track of the following states for each path:

1. The internal states of the LPC filter which are the 32 past samples of the coded speech $\hat{s}(n)$, per path.

2. The internal states of the noise-shaping section $B(z/\gamma_z)\,/\,B(z/\gamma_p)$, 16 variables per path.

3. The internal stats of the tilt section $1/T(z/\gamma_t)$. These are the 2 past samples of the shaped quantized speech $y(n)$ per path.

4. One excitation gain of previous frame per path. Recall that the gain is updated in a backward mode. Therefore, it is path dependent.

5. The internal states used in deriving the LPC filter $A(z)$ recursively [6]. There are $3(N_{lpc}+1)$ such variables where N_{lpc} is the LPC order, in our case, 99 variables per path.

The last group in the above list is of special interest. These states are needed in order to perform backward LPC analyses in all M paths using the *immediate past.* However, if one is willing to extract the LPC information from quantized speech samples earlier than L frames back, then, this information becomes common to all paths. The analysis is then, performed only once, which reduces the amount of memory and the computations needed for the LPC update, at the risk of creating some mismatch between the signal and its delayed LPC representation. However, if the LPC data varies slowly, this mismatch may be negligible.

The parameters of the noise-shaping filters $B(z)$ and $T(z)$ are derived from the input, therefore, they are common to all paths. The analysis for $B(z)$ is also done recursively but only one set of 51 variables (order 16) has to be maintained.

To perform efficiently, the system has to maintain an easily-manipulatable data structure that contains the current states, the cumulative distortions and an L-deep tree of indices. In this work, we were not concerned about the best architecture for this data structure. For the values of M used here (see next section), the size of the codebook is much greater than the width of the tree ($N \gg M$). Therefore, the assumption is that the complexity of extending the tree ($M\,N$ VQ operations) is significantly greater than that of manipulating the tree. This may not be the case if N is small [7]. The complexity of the coder is, therefore, roughly proportional to the width of the tree, namely, it is M times greater then that of the basic coder.

THE PERFORMANCE OF THE LOW-DELAY PREDICTIVE VECTOR-TREE CODER

The LDVT-CELP was simulated and tested over a 7KHz wideband speech material. As with LD-CELP, the frame was 5 samples long, the codebook was 1024 vectors in size, which, at a sampling frequency of 16000 Hz, corresponded to a bit rate of 32 Kbps. The parameters of the tree were varied in the ranges $2 \le L \le 12$ and $2 \le M \le 12$. The performance of the coder in terms of MSE SNR (dB) was recorded for each pair (M, L).

Two versions of the coder were tested. The first used path-dependent LPC analysis, namely, LPC analysis was performed for each path over the immediate past samples of the quantized speech (regular backward mode). The second coder used path-independent delayed LPC analysis, as explained in the previous section. The performance of the two versions was found to be similar for L up to about 6. For larger depths, the path-independent coder started deteriorating in comparison to the path-dependent one, probably, due to the mismatch of the delayed LPC. Therefore, we focus our interest on the path-dependent version only.

Table 1 shows the gain in SNR over the basic LD-CELP coder for the path-dependent-LPC case. The gain is shown versus the tree width M and the depth L. The end-to-end coding delay is also shown alongside the depth L. The end-to-end delay of the LDVT-CELP was computed by the formula:

$$D = \frac{(2+L)\,K}{f_s} \tag{2}$$

where f_s is the sampling frequency (16000 Hz in our case). It can be shown that this formula represents the worst-case delay value as a function of the tree depth L. For a given tree width M, the complexity of the LDVT-CELP is about M times higher than that of the basic coder. Therefore, M represents the complexity of the LDVT-CELP in terms of basic-coder complexity units. The cases $L=1$, $M=1$ are equivalent to the basic coder (gain = 0.0) and are not shown in the table.

Tree	Tree Depth L (frames) Coding Delay (msec)										
Width	2	3	4	5	6	7	8	9	10	11	12
M	1.2	1.6	1.9	2.2	2.5	2.8	3.1	3.4	3.7	4.1	4.4
2	0.79	0.61	0.75	0.53	0.66	0.67	0.65	0.69	0.72	0.75	0.80
3	0.93	0.80	0.95	1.03	0.98	0.97	1.05	1.01	1.02	1.04	1.02
4	1.16	1.32	1.42	1.50	1.42	1.45	1.46	1.49	1.50	1.52	1.50
5	1.02	1.45	1.56	1.58	1.55	1.56	1.58	1.56	1.59	1.58	1.61
6	0.88	1.28	1.30	1.48	1.63	1.62	1.54	1.58	1.63	1.67	1.68
7	1.18	1.27	1.46	1.52	1.82	1.76	1.62	1.61	1.64	1.67	1.72
8	0.89	1.57	1.62	1.69	1.70	1.79	1.77	1.76	1.78	1.78	1.81
9	1.09	1.53	1.64	1.64	1.73	1.81	1.78	1.80	1.81	1.79	1.83
10	0.97	1.54	1.62	1.65	1.76	1.84	1.85	1.84	1.82	1.91	1.99
11	0.91	1.50	1.60	1.66	1.79	1.81	1.88	1.89	2.02	2.08	2.07
12	0.75	1.41	1.48	1.73	1.86	1.80	1.91	2.04	2.10	2.32	2.19

Table 1. SNR gain of the LDVT-CELP over the LD-CELP as a function of the tree depth (delay) and the tree width (complexity).

It is intersting to observe that the gain is not a strictly monotone function of L and M, although, the general trend is clearly an increase versus L, M, as expected. The occasional decrease in the gain, may be explained by two possible effects. One is a *derailing* effect, namely, the pruning process eliminate potentially good paths and gets locked on to a locally bad path. The other effect is that of *path merging*. The M paths may become very close to one another to a point where they actually represent only one or two different paths. When this happens, the tree loses its ability to anticipate major transitions in the signal.

It should also be noticed that it does not pay to increase the delay without increasing the width, and vice versa, it does not pay to increase the width without increasing the delay. The indication is that the width (complexity) should be roughly proportional to the delay.

Table 1 shows the gain-complexity-delay tradeoff offered by the LDVT-CELP. As an example, a gain of about 1.0 dB can be achieved with a delay of only 1.2 msec and 4 complexity units, or, with a delay of 2.2 msec and 3 complexity units. A gain of 2.0 dB can be obtained with a delay of 4.4 msec and 10 complexity units, or, with a delay of 3.4 msec and 12 complexity units.

In listening tests, we have noticed that the perceptual improvement of the LDVT-CELP over the LD-CELP is noticeable even with only 1.0 dB of objective gain. This may be explained by noting that the quality of the LD-CELP is high in the first place, thus, some quantization noise is close to the threshold of hearing. That noise is pushed below the threshold in the LDVT-CELP coder.

Tree coding is only one method for handling the performance-delay-complexity tradeoffs. Future research aims at obtaining better tradeoffs possibly by other methods, with the use of perceptually-motivated, pitch-controlled, noise-shaping filters. An interesting target for this research is 16 Kbps wideband speech coding with a quality level closed to that of 32Kpbs LD-CELP or the G722-standard 64 Kbps coder.

REFERENCES

[1] Y. Shoham, Erik Ordentlich, "Low-Delay Code-Excited Linear-Predictive Coding of Wideband Speech at 32Kbps", Proc. Int. Conf. on Spoken Language Processing ICSLP-90, Nov. 1990, Vol. 1, pp. 117-120.

[2] E. Ordentlich, Y. Shoham, "Low-Delay Code-Excited Linear-Predictive Coding of Wideband Speech at 32 Kbps" Proc. ICASSP-91, pp.9-12

[3] P. Mermelstein, "G.722, a New CCITT Coding Standard for Digital Transmission of Wideband Audio Signals", IEEE Comm. Mag., pp. 8-15, Jan. 1988.

[4] CCITT Study Group XVIII, Question U/XV, Source: WP XVIII/8, "Terms of Reference of the Ad Hoc Group on 16kbit/s Speech Coding", Geneva, June 1988.

[5] J.B. Anderson, J.B. Boddie, "Tree Encoding of Speech", IEEE Trans. Inf. Theory, Vol. IT-21, pp.379-387, July 1975.

[6] T.P. Barnwell, "Recursive Windowing for Generating Autocorrelation Coefficients for LPC Analysis", IEEE Trans. ASSP, Vol. ASSP-29, No. 5, pp.1062-66, Oct. 1981.

[7] J.B. Anderson, S. Mohan, "Sequential Coding Algorithms: Survey and Cost Analysis", IEEE Trans. Comm., Vol. COM-32, No. 2, Feb. 1984.

[8] J.H. Chen, High-Quality 16Kb/s Speech Coding with a One-Way Delay Less Than 2 ms.", Proc. ICASSP-90, Vol. S1, pp. 453-6, April 1990.

18

A TWO–BAND CELP AUDIO CODER AT 16 kbit/s AND ITS EVALUATION

R. Drogo De Iacovo, R. Montagna, D. Sereno and P. Usai

Audio Coding and Transmission Quality
CSELT – Via G.Reiss Romoli, 274 – 10148 Torino ITALY

INTRODUCTION

Future applications like HDTV video, multimedia, mobile audio–visual, N–ISDN and B–ISDN services will require wideband speech coding (7 KHz bandwidth). The leading idea, supported by some preliminary experiments on human perception, is that customers will prefer systems providing images with highly natural speech and also telephone services with wideband speech.

These considerations suggest the investigation of possible schemes for high quality wideband speech coding at low bit rates (16 – 32 kbit/s) for applications like multimedia services in channels of limited capacity, or multilingual audio channels and multimedia services, leaving only a small portion of the bit rate to wideband speech, in cases with not severe channel capacity limitations.

The paper presents a possible 16 kbit/s audio (wideband speech) coding scheme and its evaluation.

CELP FEATURES

The proposed wideband speech coder is based on a two–band structure [1] in which each sub–band is coded with a CELP scheme [2] tailored to the particular sub–band. The most general scheme of a CELP coder is shown in Fig 1. In this scheme innovation, long term and short term analysis coefficients are represented by vectors belonging to suitable codebooks. Starting from this set of codebooks the combination of the three codevectors that minimize the perceptually weighted mean squared error is selected. In the split–band scheme considered in the following the two bands are obtained splitting the input signal with a QMF filter bank. The same filters recommended in G.722 standard are used.

Short term analysis, performed on the input speech with a frame duration of 15 ms, is based on the autocorrelation method for LPC coefficient evaluation and translation of LPC information to Line Spectrum Pair (LSP) domain, which allows the partitioning of spectral parameters in subsets due to the looser coupling between LSP parameters. The low sub–band (prediction order equal to 10) uses three sets, (containing the first 3 LSPs, the succeeding 3 ones and the last 4 ones) which are quantized by means of three codebooks each of them with 512 codevectors. The high sub–band (prediction order 4) uses only one set quantized by means of a codebook with 512 codevectors.

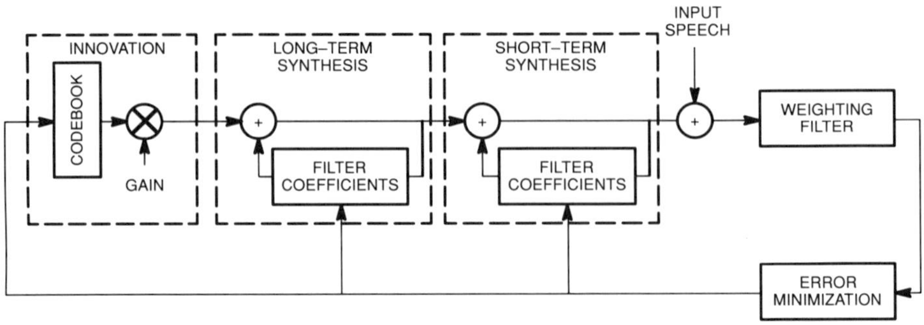

Figure 1: CELP Scheme.

In order to simplify the codec structure, Long Term (LT) analysis is used only in the low sub–band as in the high sub–band the long term correlation between two adjacent pitch periods is not very high. LT analysis is performed in closed–loop, every sub–frame of 2.5 ms duration, as described in [3]: first LT parameters are computed minimizing the squared error between the weighted input signal and the zero input response of the weighted synthesis filter, in the second step a joint optimization of the innovation signal, its gain factor and LT gain is performed. The lag is represented with 7 bits and the gain is quantized with 3 bits.

In our scheme, two different innovation codebooks are used for the two sub–bands. Both are sparse codebooks, defined starting from a limited number of codewords (keywords) identifying the pulse positions [3], the other codewords are obtained by shifting the pulses of each keyword one position at a time. A 9 bit codebook is used for the low sub–band and a 4 bit one for the high sub–band.

TWO STRATEGIES TO IMPROVE CELP PERFORMANCE

In order to improve CELP performance we investigated two possible strategies for the short–term analysis: robust linear prediction and closed–loop evaluation of the LPC coefficients.

Robust LPC Analysis

The underlying idea is that robust linear prediction [4] could improve CELP performance if sparse excitation codebooks are adopted. This is due to the characteristics of the robust linear prediction that produces an error signal with a very small variance, for the main part, and a small portion, corresponding to the excitation at the glottal openings and closures, with a larger variance. For the sake of completeness the computation steps needed [4] for the robust analysis are given in the following:

I – computation of a preliminary estimate $\underline{\tilde{a}}$ of the predictor coefficients

II – computation of the prediction residuals $\epsilon_n(\underline{\tilde{a}})$ relative to the input signal s_n

III – computation, for a given weight function $W(x)$, of the $(N\!-\!p)$ values

$$W\ (\epsilon_n(\underline{\tilde{a}}))\ \ ;\ \ n = p+1, N$$

IV – the following system of equations

$$C^{**}\underline{a} = -\underline{c}^{**}$$

with

$$C_{ij}^{**} = \sum_{n=p+1}^{N} s_{n-i}\, s_{n-j}\, W(\epsilon_n(\tilde{\underline{a}})) \; ; \; 1 \leq i,j \leq p$$

$$c_{j}^{**} = \sum_{n=p+1}^{N} s_{n-j}\, s_{n}\, W(\epsilon_n(\tilde{\underline{a}})) \; ; \; 1 \leq j \leq p$$

gives the robust parameters $\underline{a}$.

The standard method, used to obtain the preliminary estimate in the robust analysis, is the autocorrelation method. Robust linear prediction does not guarantee stability. In our simulations if an unstable filter is produced, the stable preliminary estimate is used as predictor polynomial. Less than 1% of predictors were unstable whereas about 3% were close to instability (i.e. at least one reflection coefficient module greater than 0.98).

The results of the comparison between the standard and robust analysis, when applied in the CELP scheme, are reported in Fig 2–a. It can be noted that, in spite of a slight decrease of the prediction gain, the performance in terms of SNR is enhanced. However, we found the improvement inadequate to compensate for the higher complexity of this method.

Evaluation Of LPC Coefficients In Closed–Loop

An experiment has been carried out using a binary–tree search quantization scheme in conjunction with a closed–loop evaluation of the LPC coefficients: the nearest (2^n-1) LSP–sets (to the one selected by the tree search quantization algorithm) have been used in the analysis–by–synthesis procedure and the LSP–set providing the lowest weighted error was selected and transmitted to the receiver.

	PREDICTION GAIN [dB]		CELP PERFORMANCE [dB]	
	SNR l.r.	SNR seg	SNR l.r.	SNR seg
STANDARD METHOD	12.36	14.56	11.59	11.94
ROBUST METHOD	12.21	14.45	11.74	12.36

a)

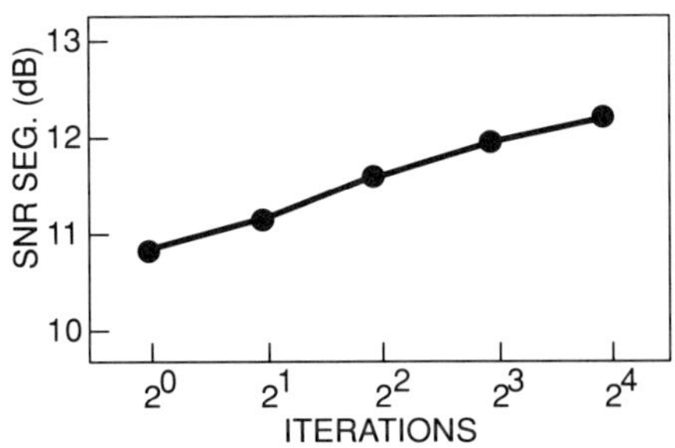

b)

Figure 2: Two Strategies To Improve CELP Performance; Comparison Between Standard And Robust LPC Analysis (a), CELP Performance With Closed–Loop Evaluation Of LPC Coefficients (b).

The results (Fig. 2–b) show that the choice between the nearest LSP–sets (n=1,2,3,4) gives improvements greater than 1 dB, in terms of CELP performance, in comparison with the standard open–loop scheme (n=0). Even though in this case the improvement is noticeable, the complexity involved is high because each LSP–set to be tested implies a complete iteration of the analysis–by–synthesis loop. Therefore, a codebook search procedure with reasonable complexity has been obtained by using a split–VQ technique and a standard open–loop scheme.

SUBJECTIVE EVALUATION

The main results obtained by means of listening tests on two speech codecs with 7 kHz bandwidth, CCITT G.722 standard (48–64 kbit/s) [5] and the described CELP scheme (16 kbit/s) are singled out. The following reference conditions were included:

- direct condition (16 bit uniform, 16 kHz sampling rate);
- PCM at 128 kbit/s (wideband codec);
- CCITT standard P.81 and a different wideband MNRU system having a noise shaping filter such that the speech quality characteristics of the latter reference system were more similar to those of the speech processed through the codec at 16 kbit/s. In the experiment the selected Q values were 5, 15, 25, 35 and 45 dB.

The average performance of two wideband MNRU reference systems, one reference codec at 128 kbit/s (PCM) and the direct condition are shown in Fig. 3–a; Fig. 3–b gives the results obtained by the two codecs having 7 kHz bandwidth, G.722 and CELP, at different speech input levels. Context effects (due to a limited number of stimulus randomizations) explain the apparent score reversal in the G.722 (Input Level=–22 dB) curve MOS vs. bit–rate.

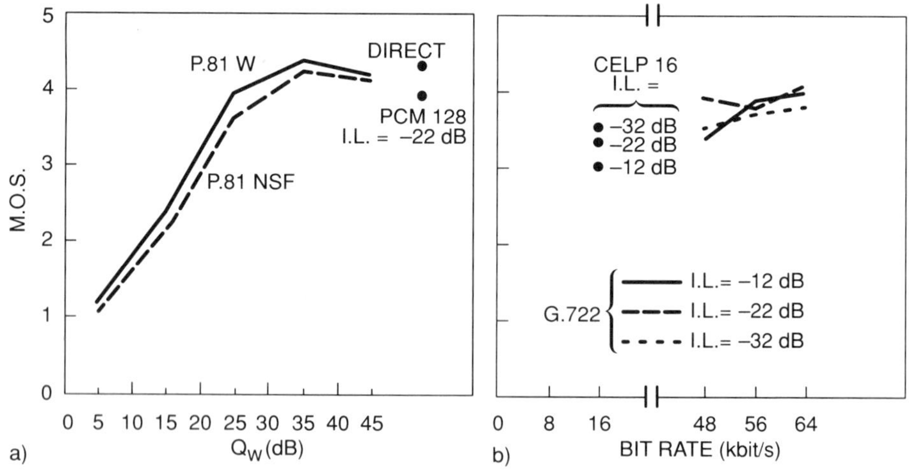

Figure 3: MOS Obtained By Reference Systems (a), MOS Obtained By G.722 And CELP Codecs At Different Input Levels (b).

The results of the experiment confirmed that the choice of the reference systems may influence the results; in fact we obtained two different behaviours with the two reference systems recommended and/or suggested by CCITT (Fig. 3–a).

Results from this experiment have also shown that the CELP codec, working at 16 kbit/s has overall quality worse than PCM at 128 kbit/s, slightly worse than G.722 at 48 kbit/s, but it sounds slightly better than narrow–band CCITT G.711 at 64 kbit/s (indirect conclusion).

CONCLUSIONS

A split–band CELP scheme, optimized to code wideband speech at 16 kbit/s, has been presented. Two different strategies to improve the coder performance have been tested: robust LPC analysis, that provides small SNRseg improvements, joint optimization of LPC coefficients and excitation parameters, that can provide up to 1 dB of SNRseg improvement. However these two strategies resulted in a considerable increase of complexity.

The subjective quality of the proposed CELP codec is slightly worse than the corresponding one of G.722 at 48 kbit/s.

REFERENCES

[1] R. Drogo De Iacovo, R. Montagna and D. Sereno, *Some experiments of 7 kHz audio coding at 16 kbit/s*, Proc. of ICASSP–89, Glasgow, pp. 192–195.

[2] B.S. Atal and M.R. Schroeder, *Stochastic coding of speech signals at very low bit rates,* Proc. Int. Conf. Commun., May 1984, part 2, pp. 1610–1613.

[3] L. Cellario, G. Ferraris and D. Sereno, *A 2–ms delay CELP coder*, Proc. of ICASSP–89, Glasgow, pp. 73–76.

[4] Chin–Hui Lee, Robust linear prediction for speech analysis, Proc. of ICASSP–87, Dallas, pp. 289–292.

[5] G. Modena, A. Coleman, P. Usai, P. Coverdale, *Subjective performance evaluation of the 7 kHz audio coder*, Proc. of GLOBECOM–86, Houston, pp. 599–602.

19

9.6 KBIT/S ACELP CODING OF WIDEBAND SPEECH

C.Laflamme, R.Salami, and J-P.Adoul

Department of Electrical Engineering,
University of Sherbrooke,
Sherbrooke, Québec, CANADA J1K 2R1

INTRODUCTION

In the recent years, there has been a great advance in the development of speech coding algorithms at very low bit rates. High-quality speech coders are now available at bit rates below 8 kb/s. Researchers' efforts, however, have focussed on narrow-band speech signals where the transmission bandwidth is limited to 300–3400 Hz, as in analog telephone systems. This bandwidth limitation degrades the speech quality, specially when the speech is to be heard through loudspeakers. For many future applications, a wider bandwidth is needed in order to achieve face-to-face communication quality. A bandwdith of 50–7000 Hz provides significantly improved quality as compared to narrow-band speech. The quality improvements are in terms of increased intelligibilty, naturalness and speaker recognition. Several future applications are foreseen for wideband speech coders, such as teleconferencing, commentary channels, and high-quality wideband telephony.

We have recently developed a high quality wideband speech coder based on algebraic code-excited linear prediction (ACELP) technique [1] which operates in the bit rate range from 9.6 to 16 kb/s. In this article, we focus on the bit rate of 9.6 kbit/s. This special bit rate will allow the transmission of AM quality speech through the present telephone lines using 9.6 kb/s modems.

As we move from narrowband to wideband, the sampling frequency is doubled, and the analysis frames will contain twice the number of samples. This will give rise to a complexity which is 3 to 4 times that of narrowband. For example, a convolution will take 4 times the number of operations, and the pitch search will take also 4 times (twice the samples and twice the delays). Further, very large excitation codebooks will be needed in order to maintain high speech quality.

In this article, we report on the strategies used to reduce the algorithmic complexity in order to allow the real time implementation of the coder on a TMS320C30 floating point processor. We also report on some procedures used to improve the speech quality at low bit rates.

ACELP STRUCTURE

In ACELP coding, a block of N speech samples is synthesized by filtering an appropriate innovation sequence from a codebook, scaled by a gain factor, through two time-varying filters. The first is known as the long-term predictor (LTP) filter, which aims at modeling the pseudo-periodicity in the speech signal (pitch periodicity). The second filter is a short-term predictor (STP) filter modeling the speech spectral envelope. It is known as the linear prediction (LP) filter and given by

$$\frac{1}{A(z)} = \frac{1}{\sum_{i=0}^{p} a_i z^{-i}}, \tag{1}$$

where p is the predictor order and a_i are the predictor coefficients. The LP coefficients are determined using the method of linear prediction analysis by minimizing the mean-square prediction error. The pitch parameters (delay and gain) and the codebook parameters (address and gain) are determined at the encoder using an analysis-by-synthesis technique. In this technique, the synthetic speech is computed for all candidate innovation sequences in the codebook retaining the particular codeword that produces the output closer to the original signal according to a perceptually weighted distortion measure.

The originality of the ACELP coder is a special innovation codebook. As opposed to most CELP coders, which utilize stochastic codewords, we use an algebraic structure that permits a fast search and requires no memory storage. Moreover, the codebook spectrum is dynamically shaped before the search procedure. This shaping utilizes a filter that emphasizes frequencies corresponding to the formants of speech resulting in an improved speech quality.

Concerning LP analysis, a 16th predictor order was found to provide the best trade-off. At 9.6 kb/s encoding rate, the filter parameters are updated every 30 ms. The LP analysis is performed using the autocorrelation method which ensures the stability of the synthesis filter. However, the increased sampling of 16000 sample/s and the higher filter order needed for wideband speech result in filters with very high prediction gains. To improve the LP analysis, two procedures were followed. The first is to preemphasize the input speech signal which has two advantages: it reduces the dynamic range of the input signal resulting in lower required precision, and it emphasizes the higher frequencies in the speech signal, so that higher frequencies can be accounted for by the transmitted excitation. The second procedure used to improve the LP analysis is to perform lag windowing on the autocorrelations of speech prior to solving the Toeplitz system of equations. Lag windowing has the effect of widening the bandwidths of the speech formants, thus avoiding bandwidth underestimation which is manifested by extremely sharp peaks in the spectral envelope. The described LP analysis was found very robust and could be implemented in single precision on the 'C30 DSP.

PITCH ANALYSIS

Pitch analysis is performed every 6 ms, and consists of determining the pitch delay and gain. The pitch parameters are usually computed in a closed loop approach, which requires filtering the previous excitation in the given delay range. The delay range 40–295 is used (8 bit adaptive codebook). The pitch delay is determined by maximizing the term

$$\mathcal{T}_\alpha = \frac{\left(\sum_{n=0}^{N-1} x(n) y_\alpha(n)\right)^2}{\sum_{n=0}^{N-1} y_\alpha^2(n)} \tag{2}$$

where $x(n)$ is the target signal given by the weighted input speech after subtracting the zero input response of the weighted synthesis filter $1/A(z/\gamma)$, and $y_\alpha(n) = u(n-\alpha) * h(n)$ is the filtered excitation at delay α ($u(n)$ is the excitation signal and $h(n)$ is the impulse response of the filter $1/A(z/\gamma)$). The complexity of the pitch search arises from the need to compute the filtered excitation $y_\alpha(n)$ for $\alpha = 40, \ldots, 295$. The convolution $u(n-\alpha) * h(n)$ can be updated exploiting the overlaping nature of the delayed excitation vectors. Using this closed loop approach, and with an optimized 'C30 code, the pitch search complexity was found to consume 120% of the real-time on the 'C30 chip. Therefore, a careful attention had to be paid to reduce the complexity of pitch computation without affecting the speech quality. The complexity reduction was accomplished using two strategies. The first is to by-pass the need to compute the filtered excitation. The second is to use decimation to reduce the number of searched delays and the number of terms in the summmations.

Eliminating the need to compute the filtered excitation can be simply done at the numerator of (2) by the use of backward filtering, whereby the numerator is given by $\sum_{n=0}^{N-1} d(n) u(n-\alpha)$ where $d(n) = x(n) * h(-n)$ is the backward filtered target signal. The denominator in (2) represents the energy of the filtered excitation. The filtered excitation need not be computed if we can find a signal which has a similar energy behaviour in the given delay range. By examining the excitation signal itself, $u(n)$, it was found to have the same energy behaviour before and after filtering. This was more evident when a stronger weighting factor of $\gamma = 0.6$ was used for pitch search. Thus the delay is now found by maximizing the term

$$\mathcal{T}_\alpha = \frac{\left(\sum_{n=0}^{N-1} d(n) u(n-\alpha)\right)^2}{\sum_{n=0}^{N-1} u^2(n-\alpha)}, \tag{3}$$

which has the complexity of an open loop approach. The correlation in (3) requires N operations and the energy can be updated using 2 operations. By taking into account the possibility of pitch multiples, the performance of this simple procedure was indistinguishable from that of the closed loop approach. The pitch search complexity in this case dropped down to 45% of the real-time.

Search method	*SNR (dB)*	% of real-time
Closed loop	17.48	120
fast approach	17.23	45
fast with decimation	17.12	20

Table 1: SNRs for different pitch search strategies.

The second approach followed to cut the complexity was by decimating the signals $d(n)$ and $u(n)$. The signals are first low-pass filtered using the single zero filter $1 + 0.7z^{-1}$ to produce the signals $u'(n)$ and $d'(n)$ given by

$$d'(n) = d(2n+1) + 0.7d(2n), \qquad n = 0, \ldots, \frac{N}{2} - 1. \tag{4}$$

Therefore, only the even values in the delay range are searched, and the number of terms in the summations in (3) is reduced to $N/2$. Once an initial even delay is determined, the two odd values around that delay are also examined and the one which minimizes the weighted error criterion is chosen. The excitation at the chosen delay is then filtered in order to determine the proper value of the gain. With this approach, we were able to cut the pitch search complexity down to 20%. This was the key factor in enabling the real-time implementation of the coder. Table 1 shows SNR values using the closed loop approach, the fast approach as in (3), and the ultra fast approach using decimation. The SNR values are averaged over 6 sentences uttered by three males and three females.

THE ALGEBRAIC CODEBOOKS

The innovation codebooks in ACELP are generated from an algebraic codebook $\mathbf{a}_k$ and a shaping matrix $\mathbf{F}$. Thus an excitation vector is given by

$$\mathbf{c}_k = \mathbf{F}\mathbf{a}_k. \tag{5}$$

The advantage of this structure is that the codebook search is decoupled from the codebook properties. The algebraic codebook is properly chosen so that it is very efficiently searched and need not be stored. The shaping matrix renders the flexibility in obtaining desired codebook properties. The search procedure can be easily brought to the algebraic domain by combining the matrix $\mathbf{F}$ with $\mathbf{H}$, the matrix containing the impulse response of the weighted synthesis filter $1/A(z/\gamma)$.

Concerning the algebraic codebooks, a two stage search using two different codebooks is performed. In the first codebook, the excitation vector contains 4

parameter	*update interval (ms)*	*bit no*
LP filter	30	54
pitch (delay and gain)	6	$8+4$
1st codebook (index and gain)	6	$12+6$
2nd codebook (index and gain)	6	$13+3$

Table 2: Bit allocation for 9.6 kb/s wideband ACELP coding.

nonzero pulses in an excitation frame of 96 samples (6 ms). The pulse amplitudes are fixed to 1, -1, 1, and -1, respectively, and their positions are given by

$$m_i^{(j)} = 3i + 12j, \qquad i = 0, \ldots, 3, \qquad j = 0, \ldots, 7. \tag{6}$$

Each pulse can have 8 possible positions distinct from the other pulses. As each pulse position is encoded with 3 bits, a 12 bit codebook is obtained. The codebook is very efficiently searched using the focussed search strategy described in [1]. The innermost loop in the search is entered 64 times at most, so that, in the worst case, only 512 codewords are examined.

As the pulse positions in the first codebook are properly chosen, the codebook is able to catch the main features in the excitation signal. The second codebook is left with almost an uncorrelated signal to model, after the pitch codebook and the first codebook have been searched. Thus a simpler model is used for the second stage codebook. A regular binary pulse excitation codebook is used [2]. A codeword contains 11 pulses with amplitudes 1 or -1 spaced by a distance of 9. The first pulse can have 4 possible positions (2 bits). This results in a 13 bit codebook. Opposed to the first codebook, in the second one the pulse position are known and we look for their optimum signs. Using the approximation that the energy of the filtered codewords is nearly constant, the pulse amplitudes are easily found as the signs of the backward target vector at the given positions.

BREAK DOWN OF THE REAL-TIME IMPLEMENTATION

The bit allocation of the 9.6 kb/s coder is shown in Table 2. The coder is implemented on a Texas Instrument TMS320C30 floating point DSP at 33.33 MHz. The processor can perform 16.7 MIPS (million instructions per second). The break-down of the percentage of real-time is shown in Table 3. It can be seen that 29% of real-time is left for codebook search. The advantage of the our codebook structure is that the search complexity can be controled using the focussed search procedure. For example, for a 'C30 DSP running at 40 MHz (now

function	*% of real-time*
Every speech frame: preemphasis, deemphasis, LPC analysis and quant.	8.3
Every subframe LSF interp., LSF to a_i, weighting $(a_i\gamma^i)$, residual	4.5
pitch search	20
compute $f(n)$, $h(n)$, correlations	13
initialize 1st book search: target vector, backward filtering	10.5
initialize 2nd book search: target vector, backward filtering	10.5
1st and 2nd book searches (controlable)	< 29
excitation, update filter memory	4.2

Table 3: Real-time break-down for the 9.6 kb/s ACELP coder.

available), we have 20% extra processing power, so we can afford to search a larger portion of the codebook. Also, with the 'C30 at 40 MHz, we can afford to implement a higher bit rate coder (say 14 kb/s) with higher quality than that at 9.6 kb/s. Although the degradation in the wideband speech at 9.6 kbs can be heard, it is vastly preferable to the original narrowband signal. Subjective tests on the 14 kbit/s version of the coder showed it is equivalent to the 56 kbit/s G.722 CCITT wideband standard.

References

[1] C.Laflamme, J-P.Adoul, R.Salami, S.Morissette, and P.Mabilleau, "16 kpbs wideband speech coding technique based on algebraic CELP" Proc. ICASSP'91, pp. 13–16.

[2] R.A.Salami, "Binary pulse excitation: a novel approach to low complexity CELP coding," in *Advances in speech coding,* pp. 145–156, B.S.Atal et al (eds), Kluwer Academic Pub., 1991.

20

HIGH FIDELITY AUDIO CODING WITH GENERALIZED PRODUCT CODE VQ[1]

Wai-Yip Chan† and Allen Gersho‡

† Department of Electrical Engineering
McGill University
Montreal, Canada H3A 2A7

‡ Department of Electrical and Computer Engineering
University of California
Santa Barbara, California 93106

INTRODUCTION

The "state-of-the-art" in wideband (e.g., 20 kHz) audio coding is exemplified by the audio component in the International Standards Organization's (ISO) Moving Picture Expert Group (MPEG) standard for the coding of audio-visual information [1]. This standard follows a basic paradigm for audio coding that prevails today. The paradigm consists of (a) variable-rate coding of samples obtained from a time-frequency analysis of the audio signal, (b) bit allocation governed by an elaborate auditory masking model to control the time-frequency distribution of quantization distortion, and (c) a constant-rate channel bit stream maintained by a buffer control loop. Virtually "transparent" compact disc (CD) quality can be obtained at 128 kb/s per channel of full 20 kHz bandwidth audio. The MPEG coding algorithm employs entropy constrained scalar quantization which can be quite efficient in terms of rate-distortion performance [2].

In this chapter, we report on the application of vector quantization (VQ) [3] to the above paradigm (with element (c) omitted) for high fidelity ("hi-fi") audio coding. VQ enjoys the optimality properties of block source coding with a fidelity criterion. However, its exponential complexity growth with block size limits the performance achievable in practice. This poor performance-complexity tradeoff can be substantially alleviated by using structured VQ methods [3] so that for a given complexity, better performance is achieved with structured than unstuctured VQ. Recently, we introduced a *generalized product code* (GPC) model that subsumes and extends most existing VQ structures and can guide the development of novel VQ structures [4]. In conjunction with the GPC model is a framework of design algorithms and paradigms for trading distortion with encoding complexity and, independently, with storage complexity, and also for the joint design of the constituent codebooks in a GPC. In the rest

[1]This work was supported in part by the the National Science Foundation, the State of California MICRO program, Rockwell International Corporation, Compression Labs, Inc., Hughes Aircraft Company, and Eastman Kodak Company.

of this chapter, we examine the quantization component of a high fidelity audio transform coder, drawing upon the GPC framework wherever appropriate. Prior descriptions of the audio coder can be found in [5] [6].

A SYNOPSIS OF THE CODING SCHEME

In our coding experiments, the audio signal is bandlimited to 15 kHz, sampled at 32 kHz, and blocked into overlapping 512-sample frames. After applying a 16 ms Hanning window to each block, corresponding to 6.7% oversampling, the discrete cosine transform (DCT) is computed. The dc coefficient is scalar quantized and 3 upper frequency coefficients are discarded. The remaining 508 transform coefficients (Figure 1) are partitioned into 127 4-dimensional *coefficient vectors* (CVs). A 127-dimensional *power envelope* vector is formed by gathering together the Euclidean norms, expressed on a dB scale, of the CVs. The power envelope vector is then quantized by the *power envelope quantizer* and the quantizer indexes are sent to the decoder. The quantized power envelope is submitted to an auditory masking computation module [7] [5] for distortion allocation among the CVs. The model specifies for each CV a target signal-to-noise ratio (SNR) that if exceeded would result in inaudible distortion, i.e. a *masking threshold* is specified in the form of an SNR value. The components of the quantized envelope are converted from dB units to linear values and applied to normalize their respective coefficient vectors. Each normalized CV is then quantized by a VQ codebook with a hybrid tree-trellis structure, with a rate necessary to meet the target SNR. The path map on the tree-trellis codebook, a variable-length descriptor, is sent to the decoder. Details concerning the quantization of the power envelope and coefficient vectors are given below.

Neither the analysis-synthesis scheme nor the perceptual distortion model employed in our experiments is as efficient as the current MPEG standard. However, the coding framework is adequate for demonstrating the effectiveness of several novel VQ structures in high- and variable-rate applications and for illustrating some aspects of the GPC design framework. Another application of GPCs, to the quantization of speech linear prediction parameters, can be found in another chapter of this volume [8].

GENERALIZED PRODUCT CODES

In product code VQ, a source is represented by $s > 1$ codebooks, one for each of s *features* $\mathbf{f}_i, i = 1, \ldots, s$. A *synthesis function* $\mathbf{g}$ maps the Cartesian product of these features into the source vector space. For a given source vector $\mathbf{x}$, the encoder selects one feature codevector $\hat{\mathbf{f}}_i, i = 1, \ldots, s$ from each codebook. Using the synthesis function, the decoder synthesizes a *reproduction* $\hat{\mathbf{x}}$, $\hat{\mathbf{x}} = \mathbf{g}(\hat{\mathbf{f}}_1, \ldots, \hat{\mathbf{f}}_s)$. In a *sequential search product code* [9], the features are encoded sequentially in s stages, from $\mathbf{f}_1$ to $\mathbf{f}_s$. In stage i, feature $\mathbf{f}_i$ is extracted from $\mathbf{x}$ via a *feature extraction rule* which may be dependent on the features quantized earlier in the sequence, i.e. $\mathbf{f}_i = \mathbf{h}(\mathbf{x}, \hat{\mathbf{f}}_1, \ldots, \hat{\mathbf{f}}_{i-1})$. The quantization

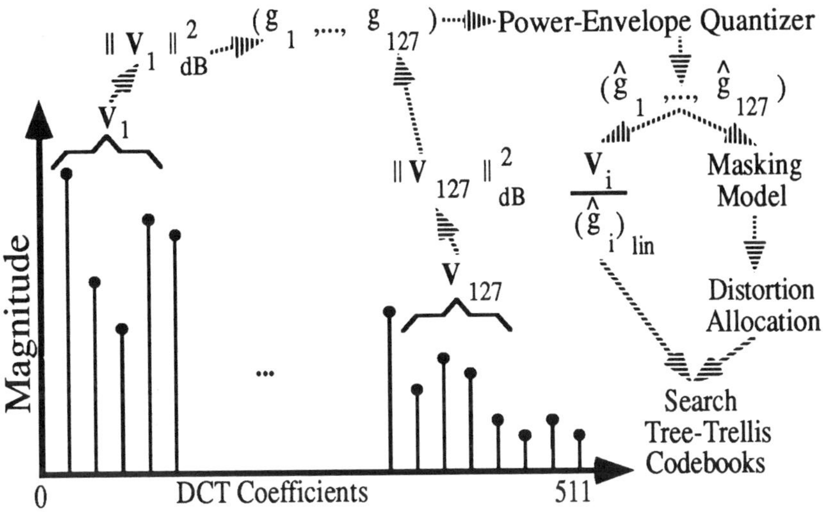

Figure 1: Feature Extraction for the Vector Quantized DCT Coefficients.

of feature $\mathbf{f}_i$ in stage i minimizes a *feature distortion measure* $d_i(\mathbf{f}_i, \hat{\mathbf{f}}_i)$ between $\mathbf{f}_i$ and all codevectors $\hat{\mathbf{f}}_i$ in the single codebook of feature $\mathbf{f}_i$. A *generalized product code* (GPC) [4] is a sequential search product code whose features $\mathbf{f}_i$, for $i > 1$, may each have $M_i \geq 1$ codebooks, $\mathcal{C}_{ij}, j = 1, \ldots, M_i$. M_i is called the *codebook fanout* for the i-th feature and M_1 is always unity. Associated with $\mathbf{c}_{ij,n}$, the nth codevector in the j-th codebook $\mathcal{C}_{ij}$ of feature $\mathbf{f}_i, i = 1, ..., s-1$, is a *codebook pointer* $\mu_{ij,n}$ whose value is the address of one of the M_{i+1} codebooks of feature $\mathbf{f}_{i+1}$. Thus, if feature $\mathbf{f}_i$ is quantized to codevector $\mathbf{c}_{ij,n}$, then the codebook addressed by the pointer $\mu_{ij,n}$ of that codevector is used for the quantization of the next feature $\mathbf{f}_{i+1}$.

The GPC structure can be graphically visualized as a trellis diagram, where the nodes are codebooks, each vertical array of nodes corresponds to a particular feature, and the directed links are the pointers that lead to the next-stage codebooks. The feature extraction rules and feature distortion measures enable the class of dynamic-programming tree search algorithms (e.g. the Viterbi and M,L algorithms) to be used to organize the search into making delayed decisions by retaining multiple evolving paths, i.e. the *survivors*, as candidates. The feature codebooks in a GPC can be designed jointly with the codebook pointers one stage at a time using the *constrained storage VQ* (CSVQ) algorithm [10]. In GPC design, the codebook fanout is treated as a storage complexity parameter, and the number of survivors kept in the delayed decision search is treated as an encoding complexity parameter. In general, the larger is either parameter, the better is the performance of the code. Moreover, the parameters can be selected independently of each other.

A product code family that is particularly relevant to our audio transform coder (and also to the work described in [8]) is the widely used *summation product codes* (SPCs). The family is defined by the prototypical synthesis function $\mathbf{x} = \mathbf{f}_1 + \cdots + \mathbf{f}_s$. The feature vectors in this family are residual vectors, obtained by subtracting from the source vector the sum of the quantized features from all previous stages, $\mathbf{f}_i = \mathbf{x} - (\hat{\mathbf{f}}_1 + \cdots + \hat{\mathbf{f}}_{i-1})$. Both the conventional multistage VQ (MSVQ) [11] and tree structured VQ (TSVQ) [12] are in this family. The difference between them is that MSVQ has unity fanout for each feature whereas TSVQ has maximum possible fanout for each feature. Split VQ (SVQ) [8] is a restricted form of MSVQ where the feature vectors have some components constrained to be zero, in such a way that the summation in the synthesis function is equivalent to concatenating subvectors and the feature extraction rule reduces to partitioning of the source vector into subvectors, one for each feature. Thus, MSVQ can outperform SVQ, though at a cost of higher complexity.

THE POWER ENVELOPE QUANTIZER

The power envelope is a very high dimensional vector with significant correlation among its components. The envelope after quantization is used to determine the masking threshold and normalize the coefficient vectors. However, an accurate rendition of such a high dimension vector using unstructured VQ would lead to an unrealizable complexity. To establish a baseline system, we first experimented with a simple envelope quantization scheme wherein the power envelope is sub-sampled to one-quarter the original frequency resolution, scalar quantized, and then interpolated onto a finer frequency grid for the purpose of normalization [5]. At 12 kb/s, we obtained 4 dB for the root-mean-square (rms) error of the interpolated envelope (whose values are in dB units).

We explored an "optimal" alternative to the above *ad hoc* interpolation scheme. In *nonlinear interpolative VQ* (NLIVQ) [13], quantization is combined with interpolation to minimize an overall distortion. NLIVQ belongs to a family of *nonlinear estimation product codes* whose prototypical synthesis function is the optimal estimator of the source vector. For the MSE distortion, the best estimator is the mean of the source vector conditioned on the features, $\mathbf{g} = E\{\mathbf{X}|\mathbf{f}_1, \ldots, \mathbf{f}_s\}$. In this application of NLIVQ, the subsampled envelope is regarded as a single feature to be quantized with unstructured or product code VQ. Interpolation is achieved by providing a conditional mean envelope vector for every quantized subsampled envelope. With only a finite number of such envelopes, the conditional means can be stored in a table; this *interpolated codebook* implements the synthesis function. The codebook size determines the rate of the NLIVQ structure, though the subsampled envelope may be quantized at the same or a higher rate. Due to the interpolated codebook, the storage complexity of the NLIVQ structure is greater than that of unstructured VQ. To circumvent this barrier and yet exploit the optimal interpolation property, we devised a two stage structure [6] in which the first stage uses NLIVQ to remove the "global" redundancy in the envelope vector. The "local" redundancy that remains is exploited in the second stage by applying SVQ to the first-stage quantization residual.

In the first stage, the power envelope is "down-sampled" to a 12-dimensional *feature power envelope*, which is then quantized with a 10-bit binary TSVQ codebook. We found that if the TSVQ codebook were replaced by a 10-bit unstructured VQ codebook, the rms error of the interpolated envelope would only improve by 0.3 dB. The index produced by quantizing the feature envelope also picks out one of the 1024 codevectors in the interpolated codebook. The interpolated envelope is then subtracted from the input envelope to obtain a residual envelope vector. By plotting the autocorrelation matrix of this residual vector, we were able to confirm that residual inter-component correlation is localized to a narrow strip along the diagonal. The residual vector is then partitioned into 13 subvectors, each to be quantized by one from a set of 6 TSVQ codebooks, with the assignment of the subvectors to the codebooks determined using the CSVQ algorithm (see next section). The resultant rms envelope error of this 2-stage scheme ranges from 2.5–4 dB at a constant bit rate of 11 kb/s. Comparing this with the projected performance of a single-stage approach in which SVQ is directly applied to the power envelope vector, we found that an additional 2 kb/s would be necessary to achieve the same level of envelope distortion; this gain is therefore attributable to the improved interpolation furnished by the NLIVQ stage.

QUANTIZATION OF THE COEFFICIENT VECTORS

Each of the 127 nominally normalized coefficient vectors has to be quantized to an SNR target specified by the masking model. A simplistic but unrealistic solution would be to provide a suite of codebooks for each vector, with each codebook in a suite designed for a target distortion level. Since the *peak* SNR necessary for distortion masking could reach almost 30 dB for some coefficient vectors, the number of codebooks would be in the thousands. On the other hand, TSVQ and all other members of the SPC family have the "successive approximation" property which can be exploited for embedded and variable rate coding. Thus, a binary TSVQ codebook designed for a peak rate of R bits/vector may also be used for encoding at all integral rates less than R. Moreover, it is not necessary to design 127 embedded codebooks for the 127 coefficient vectors. A smaller set of codebooks may be shared among the coefficient vectors in some manner such that performance is not degraded. Indeed, we determined the sharing arrangement by clustering the coefficient vector sources using the CSVQ algorithm. The clustering solutions we found admitted simple intuitive interpretations: coefficient vectors within a neighborhood on the frequency axis are grouped together; and not more than a few clusters (codebooks) suffice.

Because TSVQ is an SPC with maximum fanout, the number of codevectors grows exponentially with the number of levels. A binary TSVQ codebook that meets the peak target SNR requirement would have more than 20 levels or millions of codevectors. There is no known training algorithm that would design a balanced-tree codebook with so many nodes. Hence, we used the CSVQ algorithm to restrict the fanout of the tree as it is grown so that after say 10 levels, the fanout is kept constant and the number of nodes only grows linearly with rate. In the resultant rate-distortion characteristic, we observed that its slope in the fanout restricted portion of the tree is the same as that in the initial non-restricted portion; thus no performance penalty is paid for the storage saving and a codebook compression factor of 2-3 orders of magnitude can be obtained. Each node of the fanout-restricted tree is annotated with a SNR label. This label is acquired during the codebook training phase. In quantizing a coefficient vector, the codebook is searched as in conventional TSVQ except that at each node the target SNR of the coefficient vector is compared with the SNR label of the node. When a node is reached whose SNR label is greater than the target SNR, the search stops. A binary path map is then sent to the decoder. The decoder can determine the length of this path map while tracing out the path; the decoder has a copy of the codebook and the decoder can determine the target SNR from the quantized power envelope.

An earlier version of the audio transform coder [5] employed a rather crude envelope quantization and interpolation scheme and also *ad hoc* TSVQ codebook design to circumvent the codebook storage problem. The aforementioned improvements for the quantization of the power envelope and the coefficient vectors were able to garner savings of between 10–20 kb/s for various hi-fi audio test pieces [5]. For instance, for audio pieces on a European Broadcasting Union test CD, the average bit rate to achieve transparent quality for a piano piece is 65 kb/s, and for a guitar piece is 78 kb/s. Some very critical test pieces (e.g. a Suzzane Vega piece in the MPEG test set) however, can demand in excess of 100 kb/s for transparent quality. The performance for these are limited by other components of the coder rather than the quantizer: analysis-synthesis, masking model, and pre-echo control [1]. In any case, the gain offered by VQ over scalar quantization can still be ascertained.

CONCLUSION

We have investigated VQ of high fidelity audio signals within a perceptual-distortion constrained variable-rate transform coding framework. Structured VQ techniques are necessary to overcome the complexity barrier posed by the high quantization resolutions required to render transparent-coding quality. A prior effort using pre-existing structured VQ techniques identified several problems, two of which were the "optimal" combination of quantization with interpolation and the design of codebooks under storage constraints. We examined the problems in a larger general context of structured coder design and devised a GPC model. Associated with the model is a systematic framework for exploring and constructing structured vector quantizers. Applying some elements of this framework to enhance our earlier transform coder, we were able to obtain an average rate savings of about 15% while reducing the storage complexity by between 1–2 orders of magnitude. The resultant coder is well posed to take advantage of more efficient analysis-synthesis schemes such as those employed in the MPEG standard. In comparison with our earlier work [5], the results demonstrate that VQ offers a notable advantage over scalar quantization with scalar entropy coding and shows promise of contributing to the goal of transparent-coding quality at 64 kb/s per channel.

REFERENCES

[1] K. Brandenburg and G. Stoll, "The ISO/MPEG–Audio Codec: A Generic Standard for Coding of High Quality Digital Audio," *AES 92nd Convention*, March 1992, Reprint 3396.

[2] N. Farvardin and J.W. Modestino, "Optimum Quantizer Performance for a Class of Non-Gaussian Memoryless Sources," *IEEE Trans. Info. Th.*, pp. 485–497, May 1984.

[3] A. Gersho and R.M. Gray, *Vector Quantization and Signal Compression*, Kluwer Academic Publishers, 1992.

[4] W.Y. Chan, "The Design of Generalized Product-Code Vector Quantizers," *Proc. Int. Conf. Acoust., Sp., & Sig. Proc.*, pp. III-389–392, San Francisco, March 1992.

[5] W.Y. Chan and A. Gersho, "High Fidelity Audio Transform Coding with Vector Quantization," *Proc. Int. Conf. Acoust., Sp., & Sig. Proc.*, pp. 1109–1112, Albuquerque, April 1990.

[6] W.Y. Chan and A. Gersho, "Constrained-Storage Vector Quantization in High Fidelity Audio Transform Coding," *Proc. Int. Conf. Acoust., Sp., & Sig. Proc.*, pp. 3597–3600, Toronto, May 1991.

[7] J.D. Johnston, "Transform Coding of Audio Signals Using Perceptual Noise Criteria," *IEEE J. Sel. Areas in Comm.*, pp. 314–323, Feb. 1988.

[8] S. Wang, E. Paksoy and A. Gersho, "Product Code Vector Quantization of LPC Parameters," in this volume.

[9] W.Y. Chan and A. Gersho, "Enhanced Multistage Vector Quantization with Constrained Storage," *Proc. 24th Asilomar Conf. Cir., Sys., & Comp.*, pp. 659–663, Nov. 1990.

[10] W.Y. Chan and A. Gersho, "Constrained Storage Quantization of Multiple Vector Sources by Codebook Sharing," *IEEE Trans. Comm.*, vol. COM-38, no. 12, pp. 11–13, Jan. 1991.

[11] B.H. Juang and A.H. Gray, Jr., "Multiple Stage Vector Quantization for Speech Coding," *Proc. Int. Conf. Acoust., Sp., & Sig. Proc.*, pp. 597–600, Paris, April 1982.

[12] A. Buzo, A.H. Gray, Jr., R.M. Gray and J.D. Markel, "Speech Coding Based Upon Vector Quantization," *IEEE Trans. Acoust., Sp., & Sig. Proc.*, vol. ASSP-28, pp. 562–574, Oct. 1980.

[13] A. Gersho, "Optimal Nonlinear Interpolative Vector Quantization," *IEEE Trans. Comm.*, vol. COM-38, no. 9-10, pp. 1285–1287, Sep. 1990.

Part VI

SPEECH CODING FOR NOISY CHANNELS

Techniques for reducing the impact of channel errors on the performance of low bit rate speech coders have become very important with the use of such coders on digital cellular radio channels. The fading channels that are encountered on mobile radio systems often produce high error rates and the speech coders used on these channels must employ techniques to combat the effect of channel errors. This section reports on various aspects of source and channel coding techniques that can reduce the impact of channel errors. On radio channels, the number of bits available for forward error detection and correction is small and it is important that efficient techniques are used to protect information bits. The paper by de Marca describes a method based on unequal error protection for designing quantizers in the speech coders. Hansen et al. describe two channel coding schemes to increase the robustness in GSM half-rate channels under heavy fading conditions. The paper by Phamdo, Farvardin, and Moriya describes a low complexity robust vector quantizer (VQ) for LPS (line spectral pairs) parameters. Paliwal and Atal discuss the performance of a split VQ for LSF (line spectral frequencies) parameters in the presence of channel errors. Finally, the paper by Cox describes a generalization of the pseudo-Gray coding method of index assignment optimization for VQ codebooks.

21

ON NOISY CHANNEL QUANTIZER DESIGN FOR UNEQUAL ERROR PROTECTION

José Roberto B. de Marca*

CETUC - PUC/Rio
22453 Rio de Janeiro, RJ
BRAZIL

INTRODUCTION

In the last few years growing efforts have been devoted to greatly enhance the robustness of low bit rate speech coders to errors introduced by the transmission channel. This increasing interest is in great part due to the need for efficient coders geared towards mobile and personal communication systems. Due to the narrow spectral bandwidth assigned to these applications, the number of redundancy bits available for forward error detection and correction will necessarily be small. This will force the coder designer to use different levels of error protection. For example, the TIA standard IS-54 for cellular communications [1] incorporates three levels of protection, where only 12 out of the 159 bits in the frame are in the highest protection class while 82 are in the third class, where bits are left unprotected. The selection of which bits should be placed in which class is usually done by subjectively evaluating the impact on the received speech quality of an error in a given bit. In protection schemes like this, it is very likely that a given parameter being encoded with a B-bit quantizer will have n_1 of these bits highly protected, an average level of protection will be given to the next n_2 bits and the remaining $(B - n_1 - n_2)$ will be left unprotected. Thus, different bits of a certain binary word (index), representing a quantization level, may be subject to different error rates, due to the diverse types of channel coding being used in the protection of each bit or group of bits.

The knowledge that unequal error protection can be exploited in the design of the quantizers, vector or scalar, which are part of the speech coding system, to enhance the overall performance. In principle what is needed is to match the error protection to the error sensitivity of the different bit positions of binary word representing a given parameter. This error sensitivity can be defined as the increase in distortion when that bit position is systematically hit by a channel error.

*This work was performed while the author was on leave as a Consultant with the Speech Proc. Research Dept., AT&T Bell Laboratories, Murray Hill, NJ, U.S.A.

A tailoring of the bit error sensitivity profile can be accomplished in two ways: i) by a judicious assignment of the binary indices to the output levels, which will allow for example that errors in the more vulnerable bits will most likely cause the transmitted word to be received as one of its neighboring and; ii) by adjusting the quantizer design procedure so that the codevectors will be properly clustered.

The following two sections briefly describe the techniques available for accomplishing these two tasks. Later they are applied to the problem of waveform coding of speech and their comparative performance is exemplified.

INDEX ASSIGNMENT TECHNIQUES

An appropriate assignment of binary indices to the output levels of a quantizer can be regarded as a form of zero-redundancy channel coding. In the recent past at least three methods have been proposed [2]-[4] to solve the index assignment problem. All three algorithms are suboptimal in the sense that, in general, they will not achieve the global minimum of the distortion function. The permutation method of [2] and the simulated annealing algorithm described in [3] are search techniques which start from a random assignment and then proceed to decrease the objective function by iteratively exchanging the binary indices between two codevectors. The third scheme [4] is a constructive procedure where neighborhoods of Hamming distance 1 are sequentially built around codevectors. An attempt is made to have in those neighborhoods closest neighbors in terms of the adopted distortion function to the codevector whose index is in the center of that neighborhood. Codevectors are considered for assignment in descending order of an empirical cost function.

Let's consider a K-dimensional B-bit quantizer with codebook $\mathcal{C} = \{\mathbf{y}_i, \mathbf{y}_2, \ldots, \mathbf{y}_N\}$, where $N = 2^B$. Then the average distortion per source sample, caused by channel noise is given by [4]:

$$D_c = \frac{1}{K} \sum_{i=1}^{N} \sum_{j=1}^{N} P(\mathbf{y}_i) P(b_j|b_i) \quad d(\mathbf{y}_i, \mathbf{y}_j) \tag{1}$$

with b_i being the binary index assigned to vector $\mathbf{y}_i$ and $P(\mathbf{y}_i)$ being the a priori probability of codevector $\mathbf{y}_i$. The distortion function $d(\ ,\ )$ is some meaningful speech distortion measure, most often assumed to be a form of a weighted Euclidean measure.

All three methods just mentioned have in common that they originally proposed, assuming that the underlying channel is binary symmetric and that its crossover probability (ϵ) is small. With this assumption the only events with non-negligible probability are those with a single error per binary word. The probability $\alpha_{ij} = p(b_j|b_i)$ is then given by $\epsilon(1-\epsilon)^{B-1}$ whenever $d_h(b_i, b_j) = 1$ and zero otherwise, where $d_h(b_i, b_j)$ is the Hamming distance between indices b_i and b_j.

A byproduct of this channel choice is that the bit error sensitivity profile of the resulting binary assignment will be very close to uniform. Table 1 illustrates this fact for an 8-bit quantizer with indicies assigned by simulated annealing. The sensitivity values are normalized with respect to the sensitivity of the most significant bit (MSB) of a natural ordering assignment [5] to be described in the sequel. As can be seen, the use of the annealing algorithm results in a MSB sensitivity only twice as large as that of the LSB. On the other hand, for the Natural Ordering this factor is close to 16 and all the bits have quite different sensitivities.

It is possible however to use the simulated annealing method to obtain a non-uniform bit error sensitivity [5]. The only necessary change is in the channel model, to allow for different error rates in different bit positions. The conditional probability α_{ij} should then be expressed as:

$$\alpha_{ij} = \prod_{m=1}^{B} \epsilon_m^{\ell_m(i,j)}(1-\epsilon_m)^{1-\ell_m(i,j)} \tag{2}$$

where

$$\ell_m(i,j) = \begin{cases} 1, & \text{if } b_i \text{ and } b_j \text{ differ in position } m \\ 0, & \text{otherwise} \end{cases}$$

Index Assignment	Bits							
	MSB	7	6	5	4	3	2	LSB
Nat. Ord.	1.0	0.63	0.60	0.43	0.26	0.16	0.12	0.06
Reg. Ann.	0.44	0.42	0.40	0.40	0.35	0.31	0.24	0.22

Table 1: Error sensitivity comparison for two index assignment techniques: natural ordering and regular annealing. Numbers are normalized with respect to most significant bit (MSB) sensitivity yielded by natural ordering assignment.

The version of the annealing algorithm which incorporates eq. (2) will be referred to in this paper as channel optimized annealing, while the one which adopts the binary symmetric channel model will be called regular annealing. Table 2 illustrates the error sensitivity distribution for the COA method when the error rates affecting each bit are the following:

$$\epsilon_m = \begin{cases} 5\times10^{-4} & \text{for} \quad m = 1, 2, \ldots, S_1 \\ 5\times10^{-3} & \text{for} \quad m = S_1+1, \ldots, S_2 \\ 10^{-2} & \text{for} \quad m = S_2+1, \ldots, B \end{cases} \tag{3}$$

for $(S_1, S_2) = (2, 4)$

Clearly there is a better match to the error protection, reflected by the values of ϵ_m in (3), than obtained with the natural ordering. At this point it is worth defining what is meant by natural ordering. It is basically the result of an appropriate rearranging of the codevectors when the splitting procedure is employed for the codebook design [7]. Let's suppose, for example, that the first split has already been made producing a 1-bit codebook. The codevectors are then labeled $0(\mathbf{y}_0)$ and $1(\mathbf{y}_1)$. When a new split occurs, in principle the two new vectors could be labeled 2 and 3. This however will lead to an assignment where the bit sensitivity is not ordered. On the other hand if we label the two codevectors derived from $\mathbf{y}_0$, 0 and 1, and the ones derived from $\mathbf{y}_1$, 2 and 3, an hierarchical ordering will result. In general, after each splitting descendants of vectors labeled i should receive labels $2i$ and $2i+1$. The label (index) assignment resulting from this procedure is here called natural ordering. It is a simple and efficient way of obtaining an ordered error sensitivity distribution. It does not depend however on the values of ϵ_m and this fact will affect its performance for more complex error channel model like the one in eq. (3).

Index Assignment	Bits							
	MSB	7	6	5	4	3	2	LSB
Nat. Ord.	1.0	0.63	0.60	0.43	0.26	0.16	0.12	0.06
Ch. Opt. Ann.	0.96	0.60	0.47	0.34	0.22	0.13	0.12	0.09
COVQ	0.90	0.51	0.30	0.24	0.14	0.12	0.09	0.05
Clustering	0.80	0.50	0.35	0.23	0.15	0.10	0.09	0.05

Table 2: Error sensitivity comparison for two index assignment techniques (natural ordering and channel optimized annealing) and two VQ design procedures (channel optimized VQ and clustering). Normalization is again with respect to MSB of natural ordering.

NOISY CHANNEL CODEBOOK DESIGN

A higher degree of robustness against channel errors can be achieved if the quantizer is designed taking into account the channel induced distortion. On the other hand, although changing the index assignment does not alter the noiseless channel performance, changing the position of the output levels and decision region boundaries will worsen the behavior of the system when there are no errors.

The optimal noisy channel quantizer design criteria was introduced in [8] and basically states that codevector $\mathbf{y}_i$ should be selected to represent the input vectors s if $d^*(\mathrm{s}, \mathbf{y}_i) < d^*(\mathrm{s}, \mathbf{y}_j)$, for all $j \neq i$, $j, i = 0, 2, \ldots, 2^B - 1$, where

$d^*(\cdot\,,\,\cdot)$ is a metric which is influenced by both the speech distortion and channel model, i.e.:

$$d^*(\mathbf{s}, \mathbf{y}_i) = \sum_{j=0}^{2^B-1} \alpha_{ij}\; d(\mathbf{s}, \mathbf{y}_j)\;. \tag{4}$$

The output level $\mathbf{y}_j$, for region R_j, would then be given by:

$$\mathbf{y}_j = \sum_{i=0}^{2^B-1} \mathcal{C}(R_i)\; \alpha_{ji} \tag{5}$$

where $\mathcal{C}(R_i)$ is the centroid for region R_i.

Searching for nearest neighbor with the metric in (4) during actual quantization would be prohibitively complex. However in [9] it was shown that, at least for the Euclidean metric, it is possible to perform the search with the same level of complexity as in the noiseless LBG algorithm, without degradation in performance. It is also possible to show that the same simplification can be done for weighted Euclidean measures.

An alternative solution to the design of a noisy channel quantizer was exploited through the use of Kohonen's self-organized learning [10], [11] technique to train the quantizer. This procedure allows the clustering of the output vectors so that the pattern of distances among them has the desirable characteristics and can be summarized as follows:

i) Let $\mathbf{y}_j$ be the closest (selected) codevector to the training vector $\mathbf{s}_t$, at time t, then $\mathbf{y}_j$ is updated as:

$$\mathbf{y}_j(t+1) = (1-\alpha)\; \mathbf{y}_j(t) + \alpha \mathbf{s}_t$$

where α is an appropriately chosen adaptation rate.

ii) Let $\{\mathcal{N}_\ell, \ell = 1, \ldots, K\}$ be a set of neighborhoods of $\mathbf{y}_j$ containing each M_ℓ codevectors. Update all vectors $i^* \in \mathcal{N}_\ell$ according to

$$\mathbf{y}_{i^*}(t+1) = [1 - \epsilon_\ell(t)]\; \mathbf{y}_{i^*}(t) + \epsilon_\ell(t)\mathbf{s}_t, \quad \ell = 1, \ldots, K$$

In applying this method to VQ design for unequal error protection, the number of neighborhoods K should be made equal to $N_p - 1$, where N_p is defined as the number of protection classes. The size of a neighborhood associated with a given protection class will be a function of the number of bits included in that protection class. The set of rates $\epsilon_\ell(t)$ will control the amount of clustering which is performed and which in turn will determine the trade-off between speech distortion and channel distortion.

By using noisy channel quantizer techniques it is possible, after appropriate index assignment, to achieve extremely good matches with the protection scenario as is shown in Table 2 for the scenario in (3).

Design Method	Rate (bits/sample) 1	2
LBG	9.33	14.23
COVQ	9.25	13.66
Clustering	9.25	13.94

Table 3: Signal-to-Speech distortion ratio (SNRs) in dB for three quantization techniques for coding rates of 1 and 2 bits/samples.

Table 3 shows values of signal-to-speech distortion ratio (no channel noise) for the LBG algorithm, for a channel optimized VQ (COVQ) which was designed using eqs. (4) and (5) and for a quantizer designed using the clustering technique just described. As expected, the standard LBG algorithm provide a better performance than the noisy channel designs. The 2 bit/sample case shows that the clustering approach may be tuned to yield higher SNRs than the COVQ. However a price will have to be paid in terms of channel introduced distortion as will be seen in the next section.

RESULTS FOR WAVEFORM CODING

The methods described in the previous sections were used in the direct quantization of a speech waveform. The number of quantization levels was always 256 and two rates were considered: 1 and 2 bit/sample.

The design condition for both C.O.A and C.O.V.Q was the channel model described in eq. (3). In order to check for robustness against mismatches between the design condition and actual protection, four other scenarios were used in the evaluation. Scenario D is the design condition (no mismatch) while Scenario A assumes uniform protection. Scenarios B, C, D can be represented in terms of the parameter S_1 and S_2 defined in eq. (3): in Scenario B, $(S_1, S_2) = (1, 3)$; in C, $(S_1, S_2) = (4, 4)$ and; in E, $(S_1, S_2) = (3, 6)$.

Table 4 illustrates the behavior of three index assignment techniques for all scenarios. Results are given in terms of SNRc, the ratio between the signal power and the average expected channel introduced distortion (D_c). As expected regular annealing gives the best performance when uniform protection is adopted. But it can be more than 3 dB worse for a non-uniform protection scenario. Surprisingly, the C.O.A outperforms the natural ordering assignment (which does not depend on the actual channel) for all protection conditions. However, it is worth noting the large improvement afforded by the natural ordering.

The improvements afforded by redesigning the quantizer to account for channel distortion is exemplified in Table 5 for a 1 bit/sample quantizer. The clus-

Index Assignment	Scenario				
	D	A	B	C	E
Reg. Ann.	13.59	10.7	12.57	14.98	15.24
Nat. Ord.	14.71	9.42	12.79	18.30	16.91
COA	15.44	9.83	13.31	18.80	17.22

Table 4: Signal-to-Channel distortion ratio (SNRc) in dB for three index assignment techniques. 2 bit/sample VQ.

tering technique achieves an improvement of at least 1 dB with respect to the standard LBG algorithm, for all non-uniform scenarios. However the COVQ yields the best performance in every situation. The designer nevertheless may be interested in using the clustering approach when adopting a non-Euclidean distortion measure, which will cause the search complexity of the COVQ to be too high, or if he is willing to test or achieve different trade-offs between quantization and channel induced distortions. Both noise channel VQ design procedures appear to be very robust since their advantage over LBG is very consistent.

In summary, in this work an attempt was made to show that specially tailored index assignment and quantizer design techniques can and should be adopted so that full benefit can be obtained from non-uniform error protection schemes.

Design Method	Protection Scenario				
	D	A	B	C	E
LBG	14.47	10.03	12.95	17.16	16.48
COVQ	16.06	11.02	14.37	18.75	17.97
Clustering	15.47	10.81	13.93	18.38	17.57

Table 5: Signal-to-Channel distortion ratio (SNRc) in dB for three quantizer design procedures. Channel optimized annealing was employed for index assignment.

References

[1] Electronic Industries Association (EIA), "Cellular System," Report IS-54, December 1989.

[2] K. A. Zeger and A. Gersho, "Zero Redundancy Channel Coding in Vector Quantization," *Electronics Letters*, vol. 23, no. 12, pp. 654-656, May 1987.

[3] K. Zeger and A. Gersho, "Pseudo-Gray Coding," *IEEE Trans. on Communications*, vol. COM-38, no. 12, pp. 2147-2158, December 1990.

[4] J. Roberto B. de Marca, N. Farvardin, N. S. Jayant and Y. Shoham, "Robust Vector Quantization for Noisy Channels," *Proc. of the Mobile Satellite Conference*, pp. 515-520, Pasadena, May 1988.

[5] N. Farvardin, "A Study of Vector Quantization for Noisy Channels," *IEEE Trans. on Info. Theory*, vol. 36, pp. 799-809, July 1990.

[6] J. Roberto B. de Marca and N. S. Jayant, "An Algorithm for Assigning Binary Indices to the Codevectors of a Multi-Dimensional Quantizer," *Proc. IEEE Intl. Conf. on Communications*, pp. 1128-1132, June 1987.

[7] K. K. Paliwal and B. S. Atal, "Efficient Vector Quantization of LPC Parameters at 24 Bit/Frame," Abstracts, IEEE Workshop on Speech Coding for Telecommunications, pp. 33-35, September 1991.

[8] H. Kumasawa, M. Kasahara and T. Namekiawa, "A Construction of Vector Quantizers for Noisy Channels, *Electronics and Engineering in Japan*, vol. 67-B, no. 4, pp. 39-47, 1984.

[9] N. Farvardin and V. Vaishampayan, "On the Performance and Complexity of Channel Optimized Vector Quantizers," *IEEE Trans. on Info. Theory*, vol. 37, pp. 155-160, January 1991.

[10] A. K. Krishnamurthy *et al.*, "Neural Networks for Vector Quantization of Speech and Images," *IEEE Journal on Selected Areas in Communication*, pp. 1449-1457, October 1990.

[11] E. Yair, K. Zeger and A. Gersho, "Competitive Learning and Soft Competition for Vector Quantizer Design," *IEEE Trans. on Signal Proc.*, vol. 40, no. 2, pp. 294-309, February 1992.

22

CHANNEL CODING SCHEMES FOR THE GSM HALF-RATE SYSTEM

Henrik B. Hansen, Knud J. Larsen[1], Henrik Nielsen, and Keld B. Mikkelsen

Telecommunications Research Laboratory,
Lyngsø Allé 2, DK-2970 Hørsholm, Denmark.

[1]Institute of Circuit Theory and Telecommunication,
Building 343, Technical University of Denmark,
DK-2800 Lyngby, Denmark.

INTRODUCTION

The current activity in speech coding in Europe is focused at selecting a new pan-European standard for mobile communications known as the half-rate GSM system. The gross bit rate of a half-rate channel is limited to 11.4 kbps including both speech and channel coding, which is half the gross bit rate of the full-rate GSM traffic channel.

This chapter describes the GSM transmission channel and two channel coding schemes, which may be used to increase robustness in a heavy fading environment. The performances of the channel coding schemes, employing Reed-Solomon and punctured convolutional codes respectively, are compared in a series of experiments, and finally the impact of these results on speech quality is outlined.

CHARACTERISTICS OF THE GSM CHANNEL

The GSM system is a digital cellular mobile radio system using Time-Division-Multiplex-Access (TDMA). Each transmission frequency is shared by 8 full-rate users or up to 16 half-rate users. The transmission speed is 270 kbps and each user has access to the channel in a timeslot of 577 μs giving space for 114 bits of data and some overhead. The half-rate user is able to use this access every 10 ms by transmitting or receiving a burst of data [1]. The frame size of the speech coders described below is 20 ms, which means that a speech frame is transmitted in two bursts.

Due to multipath propagation of the radio waves, the received signal is subjected to a rather fast fading. The fading may be modelled as a Rayleigh type. The GSM-recommendation 05.05 [1] describes how to model this fading. The main extra impairment is the presence of co-channel interference from similar, but remote transmitters using the same frequency in another cell of the system. The co-channel interference is characterized by C/I, carrier-to-interference ratio.

The tests to be described in this paper are specified for the test of speech coding algorithms for the half-rate speech channel. Three test patterns (EP1, EP2, and EP3) were generated with Rayleigh fading and C/I = 10 dB, 7 dB, and 4 dB, respectively. The first two ratios correspond to 50% and 90% cell coverage while the last ratio is expected to occur outside the cell border. The reception was done by maximum-likelihood-sequence estimation using an estimate of the channel transfer function obtained from a 26 bit sequence in the overhead for each data burst. The bit error rates (BER) on the output of the maximum-likelihood receiver for EP1, EP2, and EP3 are 4.9%, 8.2%, and 13%, respectively. However, the BER does not tell the crucial fact that the errors are not randomly distributed. The errors often occur in error bursts, i.e. errors occur closely together in some data bursts. Other data bursts have only few errors. This is further accentuated by use of frequency hopping which means that the frequency is changed between each burst thus possibly reducing the correlation of fading from one burst to the other.

CHANNEL CODING SCHEMES

Two schemes may be proposed as a countermeasure against bursty errors [2]:

- Use of *bit interleaving* to spread out the errors in a nearly random fashion and then the use of a good random-error-correcting code.
- Use of *burst-error-correcting codes*, since spreading out the errors destroys valuable information about their bursty nature, which might otherwise be used by the channel decoder.

This paper considers punctured convolutional codes [3] for random error correction and Reed Solomon codes [4, 2] for burst error correction. In the following paragraphs the different properties of these codes and their applicability to the speech coding system are outlined.

Reed-Solomon (RS) codes operate on symbols with many bits, e.g. symbols with 6 bits. RS codes are well suited to combat error bursts, because closely spaced bit errors give only a few symbol errors. If the RS decoder is told that E particular symbols are in error (*erasures*) it can correct these in addition to T symbol errors at unknown positions provided that $2{\cdot}T + E \leq$ number of redundant symbols. Therefore, usage of a symbol reliability measure from the receiver may help the error-correction process. However, the reliability measure affects the whole symbol, not directly the bit(s) in error as with other decoding algorithms. This also illustrates a drawback of symbol error correction, namely that just one bit error makes the whole symbol wrong. We have tested various concatenated schemes where random bit errors are detected by a short block code, e.g. a (7,6) code. However, this uses up too much redundancy. A nice feature of RS codes is the possibility to detect some of the cases where more errors have occurred than the decoder is able to correct. The cases where the decoder detects such a situation are called decoding failures and the remaining cases with too many errors are the decoding errors. In fact, the probability of having a decoding failure in case of too many errors is $1-1/T!$ [5] and it is further possible to increase the reliability by reserving extra symbol(s) for this detection only. Thus, this inherent detection of too many errors provides an excellent basis for declaring a received frame as bad. This may be very useful for error mitigation in the speech decoder. Another strong point is the fairly low complexity of the RS decoder despite the complicated look of the algorithm [2]. The low complexity allows us to reduce the bad frame rate by decoding in a three-stage process using an increasing number of symbol erasures, as indicated in Table 2 that follows the description of the speech coder. In the first stage, the frame is decoded with few erasures, which is advantageous in frames with relatively few (random) errors, because the decoder may then locate some or all the errors itself, and at the same time reliability will be extremely high. If, however, decoding in the first stage fails, the decoder is rerun once or twice with more erasures in order to be able to utilize the full correction capacity of the decoder. In these situations, successful decoding depends very much on the ability to determine error positions using the channel state information, and the output will be less reliable.

The RS codes offer little flexibility if the encoded parameters exhibit different error sensitivities, which is usually the case in speech encoding. A definition of the requirements for reliability could be the bit error rate allowed for a certain class of protected bits. This definition should be handled with some care since, as explained above, the noise found in real mobile radio systems is bursty and the subjective effect of this noise may be very different from the subjective effect of random bit noise. Another point is that dividing symbols into different protection classes and assigning codes individually for the classes does not always improve the reliability of the most protected class, because the advantage of the greater length of a code protecting all classes may well result in the same reliability for a given amount of redundancy as the best of the short ones. This effect is certainly found for very long codes, but for the frame size used here there seems to be an advantage in dividing into different classes.

Punctured Convolutional (PC) codes are designed to correct random bit errors, and different protection classes according to specific error sensitivities are easily realized by using rate compatible puncturing [3]. The number of errors that can be corrected by a PC code is dependent on its rate (i.e. the ratio between data and the resulting code) and the so-called *memory* of the code. Low rate and/or large memory improve the error correction capability. The optimal method for decoding convolutional codes is Viterbi decoding [6, 2]. Unfortunately, the complexity of Viterbi decoding grows exponentially with the code memory. Using Viterbi decoding, the channel state information is incorporated in the decoding process in a straight-forward and efficient manner. The decoder may be augmented with an algorithm to provide a reliability measure on the output. However, this is a rather complicated task and the value of this measure may be questionable. If it is used it may provide a basis for detecting bad frames. Use of additional redundancy as a CRC may also allow such detection. This means that it is possible to trade reliability of the output data for reliability of the bad frame decision, until the scheme that best matches the requirements of the speech decoder is found. The major drawback is that the speech decoder must be able to tolerate residual bit errors in good frames in order to keep the amount of bad frames at a reasonable level. The Viterbi algorithm has the pleasant feature that the decoding errors occur in small bursts not affecting the complete speech data frame like a decoding error or a decoding failure for RS codes.

From the section on the GSM channel, it is seen that there is a considerable advantage of spreading the data in a 228 bits frame over many bursts, *burst interleaving*. However, the transmission of a speech data frame is not allowed to last longer than about 50 ms if the delay between two communicating parties should not be uncomfortable. This restriction means that a speech frame must be transmitted in 4 bursts each carrying in average 57 bits from a particular frame. The data in a 228 bits frame may thus be exposed to error bursts in one or more of the 4 bursts and in addition a few random errors may occur in the other bursts. The burst interleaving is applied in both channel coding schemes. For the PC coder, data are interleaved bit-wise to approximate a random bit error distribution. Since RS codes are burst-error-correcting, bit interleaving should not be used with the RS coder. It is, however, still desirable to have symbol error bursts distributed over several data frames, which is obtained by performing symbol-wise interleaving.

EXPERIMENTS

The channel coders are compared in terms of bit error rates, bad frame rates and complexity, and the speech quality of a 6.6 kbps and a 5.4 kbps speech coder is investigated, each of which with both Reed-Solomon and punctured convolutional coding schemes.

Speech Coding

Both speech coders used for the quality assessment are slightly modified versions of the coder presented at ICASSP-91 [7]. These speech coders are CELP-type [8] analysis-by-synthesis coders consisting of three basic functions: short-term spectrum analysis, long-term pitch prediction with fractional pitch delays, and random codebook search. The spectrum parameters are calculated once for each frame by a 10th order LPC analysis, and the LPC parameters are transformed into Line Spectrum Pairs [9] for efficient quantization. Long-term prediction is performed once per subframe by closed-loop analysis using an adaptive codebook search. In order to increase the pitch delay resolution, two adaptive codebooks are searched. The first codebook contains LPC residual sequences, thus providing whole sample delay resolution, while the second codebook contains filtered versions of these sequences to be searched for fractional sample delay vectors. In the final encoding stage, a sparse random codebook with overlapping Gaussian sequences is used.

	6.6 kbps		5.4 kbps	
Parameter	frame size	bits/frame	frame size	bits/frame
LPC Parameters:	20 ms		20 ms	
LSP $(w_1, \ldots, w_{10})$		32		33
Adaptive Codebook:	5 ms		6.67 ms	
Gain		4		4
Index		1+7		1+7
Random Codebook:	5 ms		6.67 ms	
Gain		5		5
Index		8		8

Table 1: **Bit allocations for 6.6 kbps and 5.4 kbps CELP speech coders.**

As shown in Table 1, the bit rate reduction going from 6.6 kbps to 5.4 kbps has been accomplished by reducing the number of subframes from four to three along with minor modifications in the quantization schemes. The bit error sensitivities in these coders were evaluated by means of informal listening tests, and the speech coder information bits were divided into four groups according to their perceptual importance. The adaptive codebook indices and the most significant bits of the codebook gains were found to be the most sensitive, whereas the random codebook indices are the least sensitive and in the 6.6 kbps coder these bits are transmitted without protection. The spectral parameters are fairly robust and need only moderate protection.

The speech decoder must conceal the effects of residual bit errors and bad frames as much as possible. In good frames the ordering property of the LSP parameters is checked and the order is corrected if necessary. The bad frame strategy involves partial frame substitution. The LSP's and the adaptive codebook parameters are replaced by the values of the previous frame, and the random codebook gains are faded relative to those of the previous frame. The random codebook indices are always used as they were received to avoid periodic artifacts in case of consecutive frame losses.

Performance

Table 2 summarizes the bit allocations for the PC and RS channel coders. The burst interleaving depth is 4 timeslots in all test conditions. The PC types use 4 protection classes corresponding to the perceptual importance grouping. The most significant bits located in class 3 are also protected by a CRC-check, which is used for bad frame detection. The RS types use only 2 protection classes. The bits from the groups of highest perceptual importance are merged into class 1, and the remaining bits (if any) are placed in class 0. In order to minimize

Speech Coder			6.6 kbps	5.4 kbps
coder	classes	coding rate	bits	bits
PC	PC memory	1/2	4	6
	CRC	1/2	4	4
	class 3	1/2	40	40
	class 2	1/2	8	28
	class 1	5/9	40	40
	class 0	1/1	44	0
RS	erasures (sym)		4,11,15	6,13,19
	class 1	15/31,9/19	90	108
	class 0	1/1	42	0

Table 2: **Bit allocations and protection classes per 20 ms frame for the convolutional (PC) and the Reed-Solomon (RS) channel coders.**

the bad frame rate, decoding is performed as a three-stage process using an increasing number of symbol erasures as explained in the previous section.

Speech Coder			6.6 kbps		5.4 kbps	
		without coding	PC	RS	PC	RS
BFR	EP1		3.8%	8.4%	2.4%	6.9%
	EP2		10.6%	22%	7.3%	20%
	EP3		27%	46%	21%	45%
BER	EP1	4.9%	1.71%	1.52%	0.177%	0.017%
	EP2	8.2%	2.7%	2.2%	0.58%	0.048%
	EP3	13%	4.7%	3.4%	2.2%	0.30%

Table 3: *Bad frame rate (BFR) and bit error rate (BER) of convolutional (PC) and Reed-Solomon (RS) channel coding for error patterns EP1 (C/I=10dB), EP2 (C/I=7dB), and EP3 (C/I=4dB). BER for the channel without channel coding is shown for comparison.*

A comparison between the channel coder performances for two different rates of the speech coder is shown in table 3. This table shows the output bad frame rate (BFR) which indicates the relative amount of detected bad frames, and the bit error rate (BER) measured as the total BER in good frames only. The highly reliable protection in the RS code is obtained at the expense of large bad frame rates (BFR), which are about twice those of the PC code. For the 6.6 kbps speech coder the BER is dominated by errors in the unprotected class, resulting in only a small difference between the two coding schemes. In contrast, a significant difference is found for the 5.4 kbps speech coder, and this difference is largely due to that no unprotected class is used. This makes the residual BER extremely small for the RS code.

The figures 1, 2, and 3 show details of the bit error distribution when used with the 6.6 kbps speech coder. The BER is shown for EP1, EP2, and EP3, and as in table 3, the BER is for good frames only. From the figures, it is seen that the residual BER in the protected class (bit no 42 to 131) is small for the RS code, whereas the residual BER in the protected classes (bit no 44 to 131) is significant for the PC code, especially in the EP2 and EP3 conditions. All three figures show that the error correcting capability is decreased with increased code rate for the PC code since the punctured section with rate 5/9 is found from bit no 44 to 83, and rate 1/2 is found from bit no 84 to 131. This effect may be a little exaggerated in the figures since the CRC-check applied on bit no 92 to 131 reduces the bit error rate by excluding some frames with errors in this area.

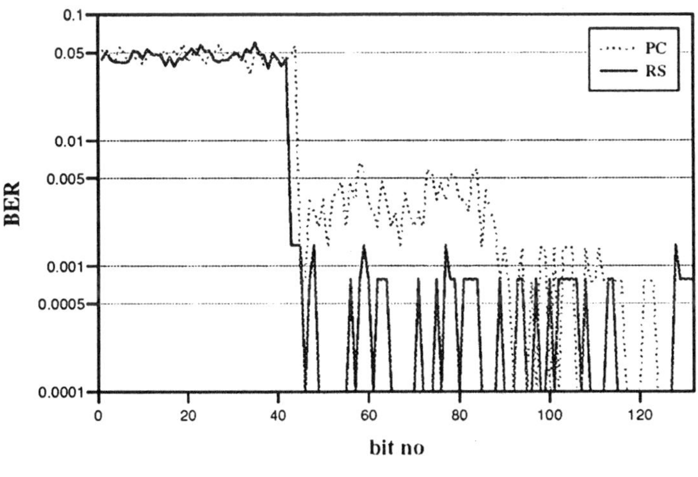

Figure 1: **Residual bit error rates of convolutional (PC) and Reed-Solomon (RS) channel coders for error pattern EP1 (C/I=10dB). Speech coder 6.6 kbps.**

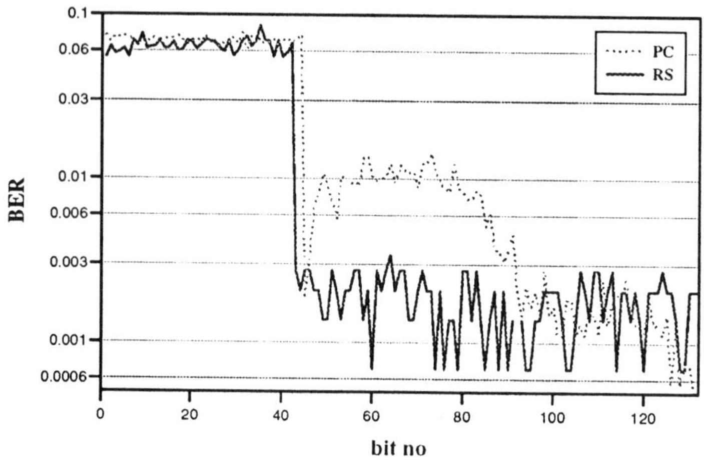

Figure 2: **Residual bit error rates of convolutional (PC) and Reed-Solomon (RS) channel coders for error pattern EP2 (C/I=7dB). Speech coder 6.6 kbps.**

CONCLUSION

The performance comparison between the PC and RS channel coding schemes shows that for the PC codes the BFR can be kept reasonably low if some residual BER is accepted, whereas the RS codes provide lower BER in exchange for significantly higher BFR.

From the performance comparison results, it is concluded that a better trade-off between residual error rate and bad frame rate is obtained using the punctured convolutional channel coders. Even though the Reed-Solomon coders provide lower residual error rate and highly reliable output, the speech quality degradation is relatively large because of the large bad frame rates.

References

[1] ETSI/GSM: Recommendations GSM 05.01, 05.03, and 05.05, ETSI 1989.

[2] G. C. Clark and J. B. Cain, *Error-Correction Coding for Digital Communications.* Plenum Press, 1988.

[3] J. Hagenauer, "Rate compatible punctured convolutional codes (RCPC-codes) and their applications," *IEEE Transactions on Communications*, vol. COM-36, pp. 389-400, 1988.

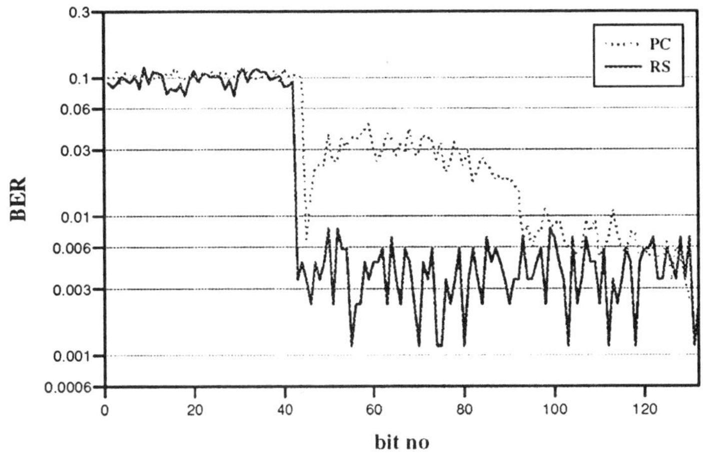

Figure 3: **Residual bit error rates of convolutional (PC) and Reed-Solomon (RS) channel coders for error pattern EP3 (C/I=4dB). Speech coder 6.6 kbps.**

[4] I. S. Reed and G. Solomon, "Polynomial codes over certain fields," *J. Soc. Ind. Appl. Math.*, vol. 8, pp. 300-304, 1960.

[5] R. J. McEliece and L. Swanson, "On the decoder error probability for Reed-Solomon codes," *IEEE Transactions on Information Theory*, vol. IT-32, pp. 701-703, 1986.

[6] A. J. Viterbi, "Error bounds for convolutional codes and an asymptotically optimum decoding algorithm," *IEEE Transactions on Information Theory*, vol. IT-13, pp. 260-269, 1967.

[7] Y. Wu, H. B. Hansen, K. J. Larsen, H. Nielsen, and J. Aa. Sørensen, "High performance coder: A possible candidate for the GSM half-rate system," in *Proc. ICASSP'91*, IEEE, 1991.

[8] M. R. Schroeder and B. S. Atal, "Code-excited linear prediction (CELP): High-quality speech at very low bit rates," in *Proc. of ICASSP'85*, pp. 937-940, IEEE, 1985.

[9] F. K. Soong and B. H. Juang, "Line Spectrum Pair (LSP) and Speech Data Compression," in *Proc. of ICASSP'84*, IEEE, 1984.

23

COMBINED SOURCE-CHANNEL CODING OF LSP PARAMETERS USING MULTI-STAGE VECTOR QUANTIZATION†

Nam Phamdo‡, Nariman Farvardin‡ and Takehiro Moriya§

‡Electrical Engineering Department
and Systems Research Center
University of Maryland
College Park, Maryland 20742

§NTT Human Interface Laboratories
Nippon Telegraph and Telephone
Corporation
3-9-11 Midori-Cho, Musashino-Shi
Tokyo, 180 Japan

INTRODUCTION

Speech coders that are robust against transmission errors are important in several applications. One specific example is in digital cellular radio [1]. In this application, the bandwidth is limited while the number of subscribers is growing; this suggests that low bit-rate (2.4 $\sim$ 4.8 kbit/s) speech coders may become more practical. At these bit rates, there is little room for error control coding. Thus, the source coding scheme to be used should be inherently robust to transmission noise.

At the rates mentioned above, the most effective coding schemes are either vocoders or hybrid coders (a combination of vocoding and waveform coding). Examples of the latter are CELP (code-excited linear prediction) [2], VSELP (vector sum excited linear prediction) [3], and TC-WVQ (transform coding with weighted vector quantization) [4]. In either case, the speech signal is separated into an excitation signal (in hybrid coders, the excitation signal is actually the output of a pitch synthesis filter) and a set of filter (or LPC) parameters, which essentially represents the short-time speech spectrum. The filter parameters play an important role in coding since a large coding error in these parameters may lead to severely degraded speech [5]. Such large errors often occur when the bit stream containing information about the filter parameters is hit with a channel error. Thus, in the design of speech coders, it is imperative that these parameters be properly encoded.

The most efficient representation of the LPC parameters is what is known as the line spectrum pair (LSP) representation [6]. In recent years, there have been numerous studies on the quantization of the LSP parameters [7-12]. In these studies, the main objective is to quantize the LSP parameters to within

†This work was supported in part by National Science Foundation grants NSFD MIP-86-57311 and NSFD CDR-85-00108, and in part by NTT Corporation and General Electric Co.

an average spectral distortion of 1 dB. In [7], [8] and [10], it was reported that about 30 to 32 bits/frame are needed when scalar quantization is used. One scheme in [9], which uses the discrete cosine transformation (DCT) and DPCM, requires 25 bits/frame while the split vector quantizer of [12] uses only 24 bits/frame. Another scheme proposed in [9] which uses 2-dimensional DCT needs just 21 bits/frame to quantize the LSP parameters, though it requires a large coding delay (100 msec). Finally, the variable-rate scheme of [11] uses just 20 bits/frame – but it is expected to be highly sensitive to transmission noise.

All of the studies mentioned above ignore the effect of channel errors on the LSP parameters. This issue will be addressed in this paper. Since channel-error propagation is not desirable, a block-structured scheme, like vector quantization (VQ), is preferred. However, ordinary VQ is not useful since it requires prohibitively large complexity in order to achieve low spectral distortion. Thus, our approach is to use multi-stage vector quantization (MSVQ), which has lower complexity than ordinary VQ. The MSVQ is matched to a channel with 1% bit error rate (BER) and the resulting scheme is called channel-matched MSVQ (CM-MSVQ). This scheme also employs a weighted squared-error distortion measure recently proposed in [13]. At 30 bits/frame, the CM-MSVQ yields an average of 0.9 dB spectral distortion in a noiseless channel. When the channel is noisy, with 1% BER, the average distortion is 1.4 dB. Comparisons are made with another scheme which basically consists of a source encoder for the LSP parameters followed by a channel encoder. Simulation results show that the CM-MSVQ scheme is superior to the other scheme over all channel conditions considered.

The remainder of this chapter is organized as follows: In the next section, a detailed discussion of the CM-MSVQ is provided. Some extensions are then briefly discussed. This is followed by a discussion on LSP parameter weighting. After that, simulation results are presented, followed by the conclusions.

CHANNEL-MATCHED MULTI-STAGE VQ

Introduction

Since Linde, Buzo and Gray (LBG) [14] provided an algorithm for its design, VQ has found many applications in the area of data compression. There are, however, two major problems associated with the LBG-VQ (also known as full-searched VQ). The first is the large complexity it requires at high bit rates and/or large block lengths. The second is its sensitivity to channel noise.

To reduce complexity, tree-structured VQ (TSVQ) [15] and MSVQ [16] had been proposed. Their performances are inferior to the ordinary VQ, but they are more practical in some situations. As to the channel-error sensitivity issue, the LBG algorithm can be modified so as to match the encoder and decoder to the channel; the resulting scheme is called channel-optimized VQ (CO-VQ) [17]. Loosely speaking, it can be said that both problems have been solved – though they have been solved independently.

In this work, we attempt to solve both of these problems jointly. That is, we seek a VQ scheme which has low complexity and at the same time is robust to channel errors. TSVQ designed for a noisy channel has been introduced in [18,19]. Here we propose a scheme called channel-matched MSVQ (CM-MSVQ).

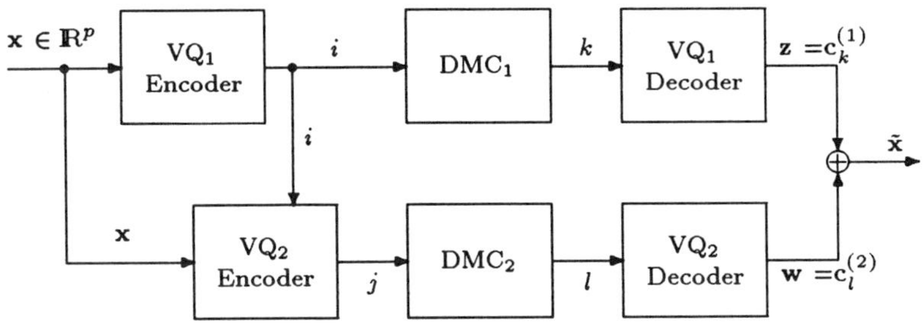

Figure 1: Block Diagram of the CM-MSVQ Scheme.

CM-MSVQ Design

Problem Statement. For the time being, let us consider only a two-stage VQ. Extension to actual multi-stage VQ will be made later on. Also, let us assume that the first-stage (or primary) VQ has already been designed and is fixed. The main problem is how to design the second-stage (or secondary) VQ given the primary VQ. Throughout, the superscripts $^{(1)}$ and $^{(2)}$ will be used to distinguish the primary and secondary VQ, respectively. Also, upper-case letters will be used to denote random variables (or vectors) while lower-case letters denote specific realizations of these random variables.

Consider the block diagram given in Figure 1. The input is assumed to be a sample of a p-dimensional random vector, $\mathbf{X}$, with probability density function $f(\mathbf{x})$. The primary VQ encoder is described by the mapping $\gamma^{(1)} : \mathbb{R}^p \longmapsto \mathcal{J}^{(1)} \triangleq \{0, 1, \ldots, N^{(1)} - 1\}$, given by

$$\gamma^{(1)}(\mathbf{x}) = i \qquad \text{if } \mathbf{x} \in S_i^{(1)}, \qquad i \in \mathcal{J}^{(1)}, \tag{1}$$

where $\mathcal{P}^{(1)} \triangleq \{S_0^{(1)}, S_1^{(1)}, \ldots, S_{N^{(1)}-1}^{(1)}\}$ is a partition of the input space, $\mathbb{R}^p$. The output of this encoder is transmitted over a discrete memoryless channel (DMC) characterized by the transition matrix $Q^{(1)}(k|i)$, where $i, k \in \mathcal{J}^{(1)}$. The primary decoder is the mapping $\beta^{(1)} : \mathcal{J}^{(1)} \longmapsto \mathbb{R}^p$, given by

$$\beta^{(1)}(k) = \mathbf{c}_k^{(1)}, \qquad \forall k \in \mathcal{J}^{(1)}, \tag{2}$$

where $\mathcal{C}^{(1)} \triangleq \{\mathbf{c}_0^{(1)}, \mathbf{c}_1^{(1)}, \ldots, \mathbf{c}_{N^{(1)}-1}^{(1)}\}$ is the primary codebook. The output of this decoder is denoted by $\mathbf{z}$. The rate of this VQ is $R^{(1)} = \frac{1}{p} \log_2 N^{(1)}$ (bits/sample).

Both the source vector, $\mathbf{x}$, and the output of the primary encoder, i, are inputs to the secondary encoder. Note that this is quite different from the noiseless-channel MSVQ, where the input to the second-stage encoder is the coding error, $\mathbf{x} - \mathbf{z}$, of the first stage. Since the channel here is noisy, the value of $\mathbf{z}$ is not known at the transmitter. Later on, we will see that what the secondary encoder actually does is encode the *expected* coding error between $\mathbf{x}$ and $\mathbf{Z}$. For now, we just assume that the secondary encoder is the mapping $\gamma^{(2)} : \mathbb{R}^p \times \mathcal{J}^{(1)} \longmapsto \mathcal{J}^{(2)} \triangleq \{0, 1, \ldots, N^{(2)} - 1\}$, described by

$$\gamma^{(2)}(\mathbf{x}, i) = j \qquad \text{if } \mathbf{x} \in S_{ij}, \qquad i \in \mathcal{J}^{(1)}, j \in \mathcal{J}^{(2)}, \tag{3}$$

where $S_{ij} \triangleq S_i^{(1)} \cap S_j^{(2)}$ and $\mathcal{P}^{(2)} \triangleq \{S_0^{(2)}, S_1^{(2)}, \ldots, S_{N^{(2)}-1}^{(2)}\}$ is another partition of $\mathbb{R}^p$. The output of this encoder is transmitted through a second DMC ($Q^{(2)}(l|j)$ where $j, l \in \mathcal{J}^{(2)}$), which is assumed to be independent from the first channel. The secondary decoder is similar in structure to the primary decoder. It is given by the mapping $\beta^{(2)} : \mathcal{J}^{(2)} \longmapsto \mathbb{R}^p$, with

$$\beta^{(2)}(l) = \mathbf{c}_l^{(2)}, \qquad \forall l \in \mathcal{J}^{(2)}, \tag{4}$$

where $\mathcal{C}^{(2)} \triangleq \{\mathbf{c}_0^{(2)}, \mathbf{c}_1^{(2)}, \ldots, \mathbf{c}_{N^{(2)}-1}^{(2)}\}$ is the secondary codebook. The output of this decoder is denoted by $\mathbf{w}$. The reconstructed vector is $\tilde{\mathbf{x}} = \mathbf{z} + \mathbf{w} = \mathbf{c}_k^{(1)} + \mathbf{c}_l^{(2)}$. The rate of the secondary VQ is $R^{(2)} = \frac{1}{p} \log_2 N^{(2)}$ (bits/sample) and the overall bit rate is $R = R^{(1)} + R^{(2)}$.

In the design problem, all parameters are fixed except for $\mathcal{P}^{(2)}$ and $\mathcal{C}^{(2)}$. Thus, the design objective is to minimize, by choices of $\mathcal{P}^{(2)}$ and $\mathcal{C}^{(2)}$, the average distortion, $D \triangleq E[d(\mathbf{X}, \tilde{\mathbf{X}})]$, where $d(\cdot, \cdot)$ is an appropriately defined distortion measure between the input vector and its reconstruction.

Solution. In [19], it was shown that D can be expressed as,

$$D = \sum_{i,j} \int_{S_{ij}} \left\{ \sum_{k,l} Q^{(1)}(k|i) Q^{(2)}(l|j) d(\mathbf{x}, \mathbf{c}_k^{(1)} + \mathbf{c}_l^{(2)}) \right\} f(\mathbf{x}) d\mathbf{x}. \tag{5}$$

The term in braces is defined as the *modified* distortion measure:

$$d_m(\mathbf{x}; i, j) \triangleq \sum_{k,l} Q^{(1)}(k|i) Q^{(2)}(l|j) d(\mathbf{x}, \mathbf{c}_k^{(1)} + \mathbf{c}_l^{(2)}). \tag{6}$$

With this definition, the CM-MSVQ design problem corresponds exactly to the MSVQ design problem for the noiseless channel, with d replaced by d_m. Specifically, for a fixed $\mathcal{C}^{(2)}$, the optimum partition $\mathcal{P}^{(2)*} \triangleq \{S_0^{(2)*}, S_1^{(2)*}, \ldots, S_{N^{(2)}-1}^{(2)*}\}$ is such that

$$S_j^{(2)*} = \left\{ \mathbf{x} : d_m(\mathbf{x}; i, j) \le d_m(\mathbf{x}; i, j') \text{ if } \mathbf{x} \in S_i^{(1)} \ \ \forall j' \in \mathcal{J}^{(2)} \right\}, \quad j \in \mathcal{J}^{(2)}. \tag{7}$$

For a fixed $\mathcal{P}^{(2)}$, the optimum codebook, $\mathcal{C}^{(2)*} \triangleq \{\mathbf{c}_0^{(2)*}, \mathbf{c}_1^{(2)*} \ldots, \mathbf{c}_{N^{(2)}-1}^{(2)*}\}$, is given by

$$\mathbf{c}_l^{(2)*} = \arg \min_{\mathbf{w} \in \mathbb{R}^p} E[d(\mathbf{X}, \tilde{\mathbf{X}})|l], \qquad l \in \mathcal{J}^{(2)}, \tag{8}$$

which can also be written as

$$\mathbf{c}_l^{(2)*} = \arg \min_{\mathbf{w} \in \mathbb{R}^p} \sum_{i,j,k} Q^{(1)}(k|i) Q^{(2)}(l|j) \int_{S_{ij}} d(\mathbf{x}, \mathbf{c}_k^{(1)} + \mathbf{w}) f(\mathbf{x}) d\mathbf{x}. \tag{9}$$

Special Case – Weighted Squared-Error. For the purpose of coding LSP parameters, we are interested in a family of distortion measures referred to as the data-dependent weighted squared-error distortion measures. Such measures are defined by

$$d(\mathbf{x}, \tilde{\mathbf{x}}) \triangleq \sum_{m=1}^{p} w_m(\mathbf{x})(x_m - \tilde{x}_m)^2, \tag{10}$$

where $\{w_m(\mathbf{x})\}_{m=1}^{p}$ is a weighting function which depends on the input vector, $\mathbf{x}$. Here, and in the subsequent discussion, we have used the generic notation x_m to denote the m-th component of the p-dimensional vector $\mathbf{x}$.

With these measures, the modified distortion measure of (6) can be simplified as follows: the term $d(\mathbf{x}, \mathbf{c}_k^{(1)} + \mathbf{c}_l^{(2)})$ can be re-written as:

$$d(\mathbf{x}, \mathbf{c}_k^{(1)} + \mathbf{c}_l^{(2)}) = \sum_{m=1}^{p} w_m(\mathbf{x}) \left[x_m^2 - 2x_m(c_{km}^{(1)} + c_{lm}^{(2)}) + (c_{km}^{(1)} + c_{lm}^{(2)})^2 \right]. \tag{11}$$

Upon defining

$$y_{im}^{(1)} \triangleq \sum_k Q^{(1)}(k|i) c_{km}^{(1)}, \qquad \forall i \in \mathcal{J}^{(1)}, \tag{12}$$

$$y_{jm}^{(2)} \triangleq \sum_l Q^{(2)}(l|j) c_{lm}^{(2)}, \qquad \forall j \in \mathcal{J}^{(2)}, \tag{13}$$

and

$$\alpha_{im}^{(1)} \triangleq \sum_k Q^{(1)}(k|i) [c_{km}^{(1)}]^2, \qquad \forall i \in \mathcal{J}^{(1)}, \tag{14}$$

$$\alpha_{jm}^{(2)} \triangleq \sum_l Q^{(2)}(l|j) [c_{lm}^{(2)}]^2, \qquad \forall j \in \mathcal{J}^{(2)}, \tag{15}$$

for $m = 1, 2, \ldots, p$, it is easy to see that the modified distortion measure of (6) is equivalent to

$$d_m(\mathbf{x}; i, j) = \sum_{m=1}^{p} w_m(\mathbf{x}) \left[x_m^2 - 2x_m(y_{im}^{(1)} + y_{jm}^{(2)}) + \alpha_{im}^{(1)} + 2y_{im}^{(1)} y_{jm}^{(2)} + \alpha_{jm}^{(2)} \right]. \tag{16}$$

The codebook search of the secondary encoder (equation (7)) involves the determination of the value of j which minimizes the above while keeping i fixed. Accordingly, for each codebook search, it suffices to compute $d_s(\mathbf{x}; i, j)$, a simplified version of $d_m(\mathbf{x}; i, j)$, which is defined as

$$d_s(\mathbf{x}; i, j) \triangleq \sum_{m=1}^{p} w_m(\mathbf{x}) \left[\alpha_{jm}^{(2)} - 2(x_m - y_{im}^{(1)})y_{jm}^{(2)}\right], \tag{17}$$

for every j in $\mathcal{J}^{(2)}$. Notice that (17) is simpler to evaluate than (16). Equation (17) corresponds exactly with the codebook search of the CO-VQ [17], with $\mathbf{x} - \mathbf{y}_i^{(1)}$ replacing $\mathbf{x}$, i.e., the secondary encoder of the CM-MSVQ acts just like the CO-VQ encoder with $\mathbf{x} - \mathbf{y}_i^{(1)}$ as the input vector. Note that $\mathbf{y}_i^{(1)}$, as defined by equation (12), is just the expected value of $\mathbf{Z}$ (the output of the primary decoder) given that i was transmitted. Hence $\mathbf{x} - \mathbf{y}_i^{(1)}$ is the expected coding error of the first-stage VQ, and it is the vector which is encoded by the second-stage VQ.

Finally, it can be readily shown that the optimum codebook of equation (9) must satisfy:

$$c_{lm}^{(2)*} = \frac{\sum_{ij} Q^{(2)}(l|j) \int_{S_{ij}} w_m(\mathbf{x})(x_m - y_{im}^{(1)}) f(\mathbf{x}) d\mathbf{x}}{\sum_{ij} Q^{(2)}(l|j) \int_{S_{ij}} w_m(\mathbf{x}) f(\mathbf{x}) d\mathbf{x}}, \tag{18}$$

for $m = 1, 2, \ldots, p$.

The second stage of the CM-MSVQ can be designed using a generalization of the LBG algorithm [14]. The complete design algorithm can be found in [19], which also includes extensive numerical results for the Gauss-Markov source.

EXTENSIONS OF CM-MSVQ

In [19], it was found that "good" results could be obtained if the first-stage quantizer is a CO-VQ designed for a channel which is *noisier* than the actual channel. It was also found that a multiple-candidate codebook search, similar to [20], provides some additional improvement at the cost of increased complexity. We have made use of these two findings in our design and implementation of the CM-MSVQ for the LSP parameters. The results will be given in a later section.

Before closing this section, let us briefly mention how the two-stage VQ can be extended to an N-stage VQ. In the design, the first $(N-1)$ stages are assumed to be given. The problem now is almost exactly the same as before, with the modified distortion measure (16) changed as follows: the term inside the brackets should be replaced by (dropping the subscript m):

$$x^2 - 2x \sum_{k=1}^{N} y_{i_k}^{(k)} + \sum_{k=1}^{N} \alpha_{i_k}^{(k)} + 2 \sum_{k=2}^{N} \sum_{l=1}^{k-1} y_{i_k}^{(k)} y_{i_l}^{(l)}, \tag{19}$$

where i_k now denotes the output of the kth-stage encoder, for $k = 1, 2, \ldots, N$. Likewise, the term in brackets of equation (17) should be replaced by

$$\alpha_{i_N}^{(N)} - 2(x - \sum_{k=1}^{N-1} y_{i_k}^{(k)}) y_{i_N}^{(N)}. \tag{20}$$

LSP PARAMETER WEIGHTING

The most commonly used objective measure for LSP parameters coding is the spectral distortion measure, given by

$$D_n = \int_{-\pi}^{\pi} (10 \log S_n(\omega) - 10 \log \tilde{S}_n(\omega))^2 \frac{d\omega}{2\pi}, \qquad (dB^2), \tag{21}$$

where $S_n(\omega)$ and $\tilde{S}_n(\omega)$ are the original and the reconstructed spectrum, respectively, associated with the n-th frame of speech and D_n is the corresponding spectral distortion. Unfortunately, there is no straightforward way of expressing the spectral distortion explicitly in terms of the LSP parameters. In most designs, either the squared-error or a weighted squared-error distortion measure is used. In this section, we introduce a weighted squared-error distortion measure, in which the weighting function depends on the difference between adjacent LSP parameters. The motivation is the following: when two adjacent LSP parameters are close to each other, this implies the presence of a spectral peak at that particular frequency. Since the locations of the spectral peaks are important in speech quality, these two parameters should be finely encoded. After trying several variations, we have decided on a weighting function which is defined by [13]:

$$w_{IHM,m}(\mathbf{x}) \triangleq \frac{1}{x_m - x_{m-1}} + \frac{1}{x_{m+1} - x_m}, \tag{22}$$

for $m = 1, 2, \ldots, p$, where x_0=0 and x_{p+1}=π. This weighting function is called inverse harmonic mean (IHM) since its inverse is the harmonic mean of the two adjacent differences. Note that when two parameters are close to each other, they will have large weights. We have found that the above weighting is effective both in terms of minimizing the average spectral distortion [13] as well as improving the perceptual speech quality.

SIMULATION RESULTS

In this section, some simulation results for quantization of LSP parameters in noisy channels are presented. We have considered three MSVQ-based schemes:

Scheme A: consists of a CM-MSVQ with five stages. Each stage is a 6-bit VQ. The first stage was designed for a channel with BER ϵ=0.1, the second stage for ϵ=0.05 and the last three stages for ϵ=0.01. This is the proposed scheme.

Scheme B: is made up of a source coder followed by a channel coder. The source coder is an MSVQ with four stages, each of which is 6-bit. The channel code is a rate-1/2 convolutional code, which only protects the six most significant bits (MSBs), i.e., the codeword of the primary VQ. The constraint length of the convolutional encoder is six (with 32 states) and the decoder is implemented using the Viterbi algorithm with the estimated codeword released at the end of each frame.

Scheme C: is a combination of Schemes A and B. It consists of a CM-MSVQ with four stages, which are exactly the same as the first four stages of Scheme A. This is followed by a rate-1/2 convolutional code which protects the six MSBs. The convolutional code is the same as in Scheme B.

In all three schemes, the IHM measure was used and the multiple-candidate codebook search was incorporated with the top four candidates passed from one stage to the next. The bit rate is 30 bits/frame for all three schemes. To simulate the channel, twenty sequences of noise were generated and the results are averaged over the 20 trials. Other experimental parameters are provided in Table 1 and the results (for inside-training and outside-training data) are given in Table 2, where we show the average spectral distortion in dB:

$$D_{ave,1} = \frac{1}{N_f} \sum_{n=1}^{N_f} (D_n)^{1/2}, \qquad (dB), \tag{23}$$

as well as in dB^2:

$$D_{ave,2} = \frac{1}{N_f} \sum_{n=1}^{N_f} D_n, \qquad (dB^2). \tag{24}$$

Here, N_f is the number of frames. The inside-training data results ($D_{ave,1}$) are also plotted in Figure 2. These results indicate that Scheme A is the best.

The complexity of this scheme is relatively low, requiring about 1.5 Mega FLOPs/sec for the codebook search, 10k words for the encoder memory and 3.1k words for the decoder memory. The proposed scheme is also robust outside the training sequence, as indicated by the results in Table 2.

Sampling Rate	8 kHz
Frame Period	22.5 *msec*
Window	30 *msec* Hamming
Analysis Order	10
Training Frames	27,968
Testing Frames	1232 (inside) 2261 (outside)
Number of Tests	20

Table 1: Experimental Conditions.

CONCLUSIONS

An efficient and robust coding scheme for speech LSP parameters has been proposed. This scheme is a channel-matched multi-stage VQ with a multiple-candidate codebook search using a weighted squared-error distortion measure.

	Bit Error Rates				
	ϵ=0.00	ϵ=0.005	ϵ=0.01	ϵ=0.05	ϵ=0.10
Inside-Training Data					
Scheme A	0.92 (0.94)	1.16 (1.96)	1.38 (2.98)	2.83 (10.96)	4.09 (20.61)
Scheme B	1.18 (1.51)	1.31 (2.26)	1.47 (3.28)	3.14 (17.94)	5.91 (52.80)
Scheme C	1.30 (1.84)	1.39 (2.29)	1.51 (2.97)	2.84 (12.99)	5.07 (37.04)
Outside-Training Data					
Scheme A	0.94 (0.97)	1.17 (1.98)	1.40 (3.04)	2.87 (11.30)	4.14 (21.15)
Scheme B	1.21 (1.58)	1.36 (2.37)	1.52 (3.36)	3.21 (18.74)	6.07 (55.33)
Scheme C	1.29 (1.79)	1.42 (2.40)	1.56 (3.15)	2.95 (13.92)	5.28 (39.43)

Table 2: Average Spectral Distortion for Inside-Training and Outside-Training Data; Values Given Are $D_{ave,1}$ in dB; Values in Parentheses Are $D_{ave,2}$ in dB^2; Rate is 30 Bits/Frame.

It has relatively low complexity and is also robust to data outside the training sequence. A possible application of this scheme is in low bit-rate digital mobile radio.

It should be mentioned that the interframe correlation of the LSP parameters had not been exploited by the CM-MSVQ. If this property is utilized, additional improvements may be obtained [21]-[23].

REFERENCES

[1] M. J. McLaughlin and P. D. Rasky, "Speech and Channel Coding for Digital Land-Mobile Radio," *IEEE Journal on Selected Areas in Communications*, vol. 6, pp. 332-344, Feb. 1988.

[2] M. S. Schroeder and B. S. Atal, "Code-Excited Linear Prediction (CELP): High Quality Speech at Very Low Bit Rates," *Proc. ICASSP-85*, pp. 937-940.

[3] G. Davidson and A. Gersho, "Complexity Reduction Methods for Vector Excitation Coding," *Proc. ICASSP-86*, pp. 3055-3058.

[4] T. Moriya and M. Honda, "Transform Coding of Speech Using a Weighted Vector Quantizer," *IEEE Journal on Selected Areas in Communications*, vol. 6, pp. 425-431, Feb. 1988.

[5] T. Moriya and H. Suda, "An 8 kbit/s Transform Coder for Noisy Channel," *Proc. ICASSP-89*, pp. 325-328.

[6] N. Sugamura and F. Itakura, "Speech Data Compression by LSP Speech Analysis-Synthesis Technique," *IECE Trans.*, vol. J64-A, No. 8, pp. 599-605, Aug. 1981 (in Japanese).

[7] N. Sugamura and N. Farvardin, "Quantizer Design in LSP Speech Analysis-Synthesis," *IEEE Journal on Selected Areas in Communications*, vol. 6, pp. 432-440, Feb. 1988.

[8] F. K. Soong and B. H. Juang, "Optimal Quantization of LSP Parameters," *Proc. ICASSP-88*, pp. 394-397.

[9] N. Farvardin and R. Laroia, "Efficient Encoding of Speech LSP Parameters Using the Discrete Cosine Transformation," *Proc. ICASSP-89*, pp. 168-171.

[10] F. K. Soong and B. H. Juang, "Optimal Quantization of LSP Parameters Using Delayed Decisions," *Proc. ICASSP-90*, pp. 185-188.

[11] N. Phamdo and N. Farvardin, "Coding of Speech LSP Parameters Using TSVQ with Interblock Noiseless Coding," *Proc. ICASSP-90*, pp. 189-192.

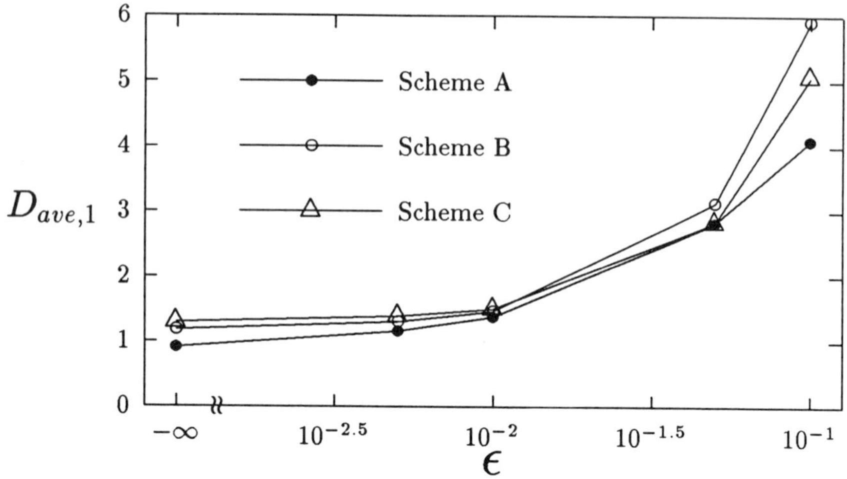

Figure 2: Simulation Results for Three Schemes (in dB) for Inside-Training Data; Rate is 30 Bits/Frame.

[12] K. K. Paliwal and B. S. Atal, "Efficient Vector Quantization of LPC Parameters at 24 bits/frame," *Proc. ICASSP-91*, pp. 661-664.

[13] R. Laroia, N. Phamdo and N. Farvardin, "Robust and Efficient Quantization of Speech LSP Parameters Using Structured Vector Quantization," *Proc. ICASSP-91*, pp. 641-644.

[14] Y. Linde, A. Buzo, and R. M. Gray, "An Algorithm for Vector Quantization Design," *IEEE Trans. Commun.*, vol. 28, pp. 84-95, Dec. 1980.

[15] R. M. Gray and Y. Linde, "Vector Quantizers and Predictive Quantizers for Gauss-Markov Sources," *IEEE Trans. Commun.*, vol. COM-30, pp. 381-389, Feb. 1982.

[16] B. H. Juang and A. H. Gray, Jr., "Multiple Stage Vector Quantization for Speech Coding," *Proc. ICASSP-82*, pp. 597-600.

[17] N. Farvardin and V. Vaishampayan, "On the Performance and Complexity of Channel-Optimized Vector Quantizers," *IEEE Trans. Inform. Theory,* pp. 155-160, Jan. 1991.

[18] N. Phamdo, "Coding of Speech LSP Parameters Using Tree-Searched Vector Quantization," M.S. Thesis, *University of Maryland*, 1989.

[19] N. Phamdo, N. Farvardin and T. Moriya, "A Unified Approach to Tree-Structured and Multi-Stage Vector Quantization for Noisy Channels," accepted for publication in *IEEE Trans. Inform. Theory,* Jul. 1992.

[20] W. LeBlanc, S. Mahmoud and V. Cuperman, "Joint Design of Multi-Stage VQ Codebooks for LSP Quantization with Applications to 4 kb/s Speech Coding," *in this book.*

[21] N. Phamdo, N. Farvardin and T. Moriya, "Channel-Error Protection for LSP Parameters Using the Interframe Correlation Property," *Proc. of the 1990 Autumn Meeting of the Acoustical Society of Japan,* pp. 193-194, Sept. 1990.

[22] N. Phamdo and N. Farvardin, "Optimal Detection of Discrete Markov Sources Over Discrete Memoryless Channels — Applications to Combined Source-Channel Coding," submitted to *IEEE Trans. Inform. Theory,* Mar. 1992.

[23] Y. Hussain and N. Farvardin, "Finite-State Vector Quantization for Noisy Channels," submitted to *IEEE Trans. Sig. Proc.,* Jun. 1992.

24

VECTOR QUANTIZATION OF LPC PARAMETERS IN THE PRESENCE OF CHANNEL ERRORS

K.K. Paliwal and B.S. Atal

AT&T Bell Laboratories
Murray Hill, New Jersey 07974, USA

INTRODUCTION

Linear predictive coding (LPC) parameters are widely used in various speech coding applications for representing the short-time spectral envelope information of speech [1]. For low bit rate speech coding applications, it is important to quantize these parameters using as few bits as possible. Considerable work has been done in the past to develop both scalar and vector quantization procedures to quantize the LPC parameters [2, 3, 4]. Scalar quantizers quantize each of the LPC parameters independently, while vector quantizers consider the entire set of LPC parameters as an entity and allow for direct minimization of quantization distortion. Because of this, the vector quantizers result in smaller distortion than the scalar quantizers at any given bit rate. The vector quantizers, however, have one major problem; their computational complexity is high. In our earlier paper [3], we have reported on a vector quantizer where the LPC parameter vector is split in the line spectral frequency (LSF) domain to overcome this complexity problem. We have shown that this quantizer can quantize the LPC parameters at 24 bits/frame with an average spectral distortion of 1 dB, less than 2% frames having spectral distortion [1] in the range 2-4 dB and no frame having spectral distortion greater than 4 dB.

In this paper, we study the performance of this vector quantizer in the presence of channel errors and compare it with that of the scalar quantizers. We also investigate the use of error correcting codes for improving the performance of the vector quantizer in the presence of channel errors.

VECTOR QUANTIZATION OF LPC PARAMETERS

In this section, we describe the split vector quantizer which is used in this paper to investigate the effect of channel errors. In [3], we have shown that the LSF representation is better suited for split vector quantization of LPC parameters than the other representations such as the log-area ratio and arcsine reflection coefficient representations. For vector quantization of LPC parameters, we divide the LSF vector into two parts. For minimizing the complexity of the split vector quantizer, the total number of bits available for LPC quantization are divided equally to the individual parts. Thus, for a 24 bits/frame LPC quantizer, each of the two parts is allocated 12 bits.

[1] In this paper, LPC quantization results are reported for telephone speech, digitized at 8 kHz sampling rate and analyzed using a 10th-order LPC analysis.

Selection of a proper distortion measure is the most important issue in the design and operation of a vector quantizer. In [3], we proposed a weighted Euclidean distance measure for this purpose and showed that it offers an advantage of about 2 bits/frame over the conventional Euclidean distance measure. The weighted Euclidean distance measure $d(\mathbf{f}, \hat{\mathbf{f}})$ between the test LSF vector $\mathbf{f}$ and the reference LSF vector $\hat{\mathbf{f}}$ is given by

$$d(\mathbf{f}, \hat{\mathbf{f}}) = \sum_{i=1}^{10} [w_i(f_i - \hat{f}_i)]^2, \tag{1}$$

where f_i and $\hat{f}_i$ are the i-th LSFs in the test and reference vector, respectively, and w_i is the weight assigned to the i-th LSF. It is given by

$$w_i = [P(f_i)]^r, \tag{2}$$

where $P(f)$ is the LPC power spectrum associated with the test vector as a function of frequency f and r is an empirical constant which controls the relative weights given to different LSFs and is determined experimentally. A value of r equal to 0.15 has been found satisfactory.

In the weighted Euclidean distance measure, the weight assigned to a given LSF is proportional to the value of the LPC power spectrum at the LSF. Thus, this distance measure allows for quantization of LSFs in the formant regions better than those in the non-formant regions. Also, the distance measure gives more weight to the LSFs corresponding to the high-amplitude formants than to those corresponding to the lower-amplitude formants; the LSFs corresponding to the valleys in the LPC spectrum get the least weight. We have used this distance measure earlier for speech recognition and obtained good results [5].

It is well known that the human ear cannot resolve differences at high frequencies as accurately as at low frequencies. We, therefore, give more weight to the lower LSFs than to the higher LSFs and modify the distance measure by introducing an additional weighting term as follows:

$$d(\mathbf{f}, \hat{\mathbf{f}}) = \sum_{i=1}^{10} [c_i w_i(f_i - \hat{f}_i)]^2, \tag{3}$$

where c_i is the additional weight assigned to the i-th LSF. In the present study, the values of $\{c_i\}$ are experimentally determined. The following values are found to be satisfactory:

$$c_i = \begin{cases} 1.0, & \text{for } 1 \le i \le 8, \\ 0.8, & \text{for } i = 9, \\ 0.4 & \text{for } i = 10. \end{cases} \tag{4}$$

Note that in (3), the weights $\{w_i\}$ vary from frame-to-frame depending on the LPC power spectrum, while the weights $\{c_i\}$ do not change from frame-to-frame. We call the distance measure defined by (3) as the weighted LSF distance measure.

In order to study the performance of this vector quantizer, we use a speech data base consisting of 23 minutes of speech recorded from 35 different FM radio stations. The first 1200 s of speech (from about 170 speakers) is used for training, and the last 160 s of speech (from 25 speakers, different from those used for training) is used for testing. Speech is lowpass filtered at 3.4 kHz and digitized at a sampling rate of 8 kHz. A tenth order LPC analysis, based on the stabilized covariance method with high frequency compensation [6] and error weighting [7], is performed every 20 ms using a 20-ms analysis window. Thus, we have here 60,000 LPC vectors for training, and 8000 LPC vectors for testing. We will refer to this data base as the 'FM radio' data base. In order to avoid sharp spectral peaks in the LPC spectrum which may result in unnatural synthesized speech, a fixed bandwidth expansion of 10 Hz is applied to each pole of the LPC vector, by replacing the predictor coefficient a_i by $a_i\gamma^i$, for $1 \leq i \leq 10$, where $\gamma = 0.996$.

The split vector quantizer with the weighted LSF distance measure is studied at different bit rates. Spectral distortion (defined as the root mean square difference between the original LPC log-power spectrum and the quantized LPC log-power spectrum) is used as a criterion for evaluating the LPC quantization performance. Results are shown in Table 1. We can see from this table that we need only 24 bits/frame to get "transparent" quality LPC quantization. (By "transparent" quantization of LPC information, we mean that the LPC quantization does not introduce any additional audible distortion in the coded speech, i.e., the two versions of coded speech – the one obtained by using unquantized LPC parameters and the other by using the quantized LPC parameters – are indistinguishable through listening. It is generally agreed [2, 3] that transparent quantization of LPC information can be obtained by maintaining the following three conditions: 1) the average spectral distortion is about 1 dB, 2) there is no outlier frame having spectral distortion larger than 4 dB, and 3) the number of outlier frames having spectral distortion in the range 2-4 dB is less than 2%.)

Bits used	Av. SD (in dB)	Outliers (in %)	
		2-4 dB	>4 dB
26	0.90	0.44	0.00
25	0.96	0.61	0.00
24	1.03	1.03	0.00
23	1.10	1.60	0.00
22	1.17	2.73	0.00
21	1.27	4.70	0.00
20	1.34	6.35	0.00

Table 1. Spectral distortion (SD) performance of the split vector quantizer as a function of bit rate using the weighted LSF distance measure.

In order to put this quantizer in proper perspective, we compare its performance with that of the optimal nonuniform scalar quantizers which are designed here for the following LPC parameters: 1) the LSFs, 2) the LSF differences, 3) the arcsine reflection coefficients, and 4) the log-area ratios. These quantizers are designed by using the LBG algorithm [8] on the training data. Different number of bits are used to quantize each LPC parameter. Nonuniform bit allocation is determined from the training data using a method described in [9]. The LPC quantization performance of each of these quantizers is listed in Table 2 for different bit rates. By comparing this table with Table 1, we can

Bits used	Para-meter	Av. SD (in dB)	Outliers (in %)	
			2-4 dB	>4 dB
36	LSF	0.79	0.46	0.00
36	LSFD	0.75	0.60	0.01
36	ASRC	0.81	0.90	0.01
36	LAR	0.80	1.09	0.04
34	LSF	0.92	1.00	0.01
34	LSFD	0.86	1.10	0.01
34	ASRC	0.92	2.05	0.08
34	LAR	0.92	1.65	0.04
32	LSF	1.10	2.21	0.03
32	LSFD	1.05	3.13	0.01
32	ASRC	1.04	3.30	0.09
32	LAR	1.04	3.20	0.04
28	LSF	1.40	9.21	0.05
28	LSFD	1.25	7.36	0.05
28	ASRC	1.32	9.29	0.23
28	LAR	1.34	9.51	0.16

Table 2. Spectral distortion (SD) performance of different scalar quantizers using the LSF, LSF difference (LSFD), arcsine reflection coefficient (ASRC) and log-area ratio (LAR) representations.

see that the 24 bits/frame split vector quantizer is comparable in performance with the scalar quantizers operating at bit rates in the range 32-36 bits/frame. We also compare the 24 bits/frame split vector quantizer with the 34 bits/frame LSF scalar quantizer used in the U.S. federal standard 4.8 kb/s code-excited linear prediction (CELP) coder [10]. This scalar quantizer (to be called LSF-FS) results in average spectral distortion of 1.45 dB, 11.16% outliers in the range 2-4 dB, and 0.01% outliers having spectral distortion greater than 4 dB. It is clear that the 24 bits/frame split vector quantizer performs better than the 34 bits/frame LSF scalar quantizer used in the federal standard 4.8 kb/s CELP coder.

EFFECT OF CHANNEL ERRORS

In the preceding sections, we have shown that the split vector quantizer can quantize LPC information with transparent quality using 24 bits/frame. In order to be useful in a practical communication system, this quantizer should be able to cope with the channel errors. In this section, we study the performance of this quantizer in the presence of channel errors and compare it with that of the scalar quantizers. We also investigate the use of error correcting codes for improving the performance of the split vector quantizer in the presence of channel errors.

Channel errors, if not dealt with properly, can cause a significant degradation in the performance of a vector quantizer. This problem has been addressed recently in a number of studies [11, 12, 13], where algorithms for designing a quantizer that is robust in the presence of channel errors were described. In these robust design algorithms, the codebook is reordered (or, the codevector indices are permuted) such that the Hamming distance between any two codevector indices corresponds closely to the Euclidean distance between the corresponding codevectors. Farvardin [12] has used the simulated annealing algorithm to design such a codebook. However, he has observed that when the splitting method [8] is used for the initialization of the vector quantizer design algorithm, the resulting codebook has a "natural" ordering which is as good in the presence of channel errors as that obtained by using the simulated annealing algorithm, especially for sources with memory (i.e., where vector-components are correlated). In our experiments with the split vector quantizer, we have made similar observations. Since the naturally-ordered codebook is obtained without additional computational effort and it performs well in the presence of channel errors, we use it in our experiments. Naturally-ordered codevectors in this codebook have the property that the most significant bits of their binary addresses are more sensitive to channel errors than the least significant bits, i.e., a channel error in the most significant bit in the binary address of a codevector causes a larger distortion than that in the least significant bit. In our experiments described in this section, we use this property to our advantage by protecting the most significant bits by using error correcting codes.

Performance of the 24 bits/frame split vector quantizer is studied for different bit error rates and results (in terms of spectral distortion) are shown in Table 3. Naturally-

Bit error rate (in %)	Av. SD (in dB)	Outliers (in %)	
		2-4 dB	>4 dB
0.0	1.03	1.03	0.00
0.001	1.03	1.04	0.01
0.01	1.03	1.09	0.04
0.05	1.05	1.41	0.30
0.1	1.08	2.00	0.64
0.5	1.28	5.55	3.11
1.0	1.55	9.73	6.76
10.0	4.62	27.68	54.69

Table 3. Effect of channel errors on the spectral distortion (SD) performance of the 24 bits/frame split vector quantizer.

ordered codebooks (obtained by using the splitting method for the initialization of the vector quantizer design algorithm) are used in this study. It can be seen from Table 3 that the channel errors result in outlier frames having spectral distortion greater than 4 dB, even for a bit error rate as small as 0.001%. Thus, the split vector quantizer does not have transparent quality in the presence of channel errors. However, it results in an average spectral distortion of about 1 dB for a bit error rate as high as 0.1%.

In order to put the performance of the split vector quantizer in proper perspective, we study here the effect of channel errors on the performance of the following two 34 bits/frame scalar quantizers: one using LSFs and the other using log-area ratios. Results (in terms of spectral distortion) for these two quantizers for different bit error rates are shown in Tables 4 and 5, respectively. Note that the 34 bits/frame LSF-based

Bit error rate (in %)	Av. SD (in dB)	Outliers (in %)	
		2-4 dB	>4 dB
0.0	0.92	1.00	0.01
0.001	0.92	1.01	0.03
0.01	0.93	1.09	0.11
0.05	0.95	1.51	0.36
0.1	0.98	1.96	0.80
0.5	1.23	5.56	4.01
1.0	1.56	9.35	8.38
10.0	5.12	23.30	62.25

Table 4. Effect of channel errors on the spectral distortion (SD) performance of the 34 bits/frame LSF-based (LSF-FS) scalar quantizer.

Bit error rate (in %)	Av. SD (in dB)	Outliers (in %)	
		2-4 dB	>4 dB
0.0	0.92	1.65	0.04
0.001	0.92	1.65	0.06
0.01	0.93	1.69	0.13
0.05	0.95	1.99	0.38
0.1	0.99	2.60	0.65
0.5	1.25	7.10	3.30
1.0	1.55	12.44	6.21
10.0	5.38	27.99	58.89

Table 5. Effect of channel errors on the spectral distortion (SD) performance of the 34 bits/frame log-area ratio based scalar quantizer.

scalar quantizer has been used in the U.S. federal standard CELP coder [10] because it was found to be quite robust to channel errors and its performance degraded gracefully for larger bit error rates. By comparing Tables 4 and 5 with Table 3, we can observe that, like the 24 bits/frame split vector quantizer, the 34 bits/frame scalar quantizers are unable to attain transparent quality in the presence of channel errors for a bit error rate as small as 0.001%. Also, both the scalar quantizers can provide an average spectral distortion of about 1 dB with a bit error rate of 0.1%. For larger bit error rates, the scalar quantizers show more degradation in performance than the split vector quantizer. Thus, the 24 bits/frame split vector quantizer compares favorably with respect to the 34 bits/frame scalar quantizers in terms of its performance in the presence of channel errors.

So far, the effect of channel errors on the performance of the LPC quantizers has been studied in terms of spectral distortion. Now, we study how the distortion due to channel errors affects the quality of the synthesized speech from a given coder. For this, we use a CELP coder[2] and assume that the channel errors affect only the LPC parameters. Here, we use a database consisting of 48 English sentences spoken by 6 male and 6 female speakers. These sentences are processed by the CELP coder and segmental signal-to-noise ratio of the coded speech is computed for different bit error rates. Results are shown in Table 6 for the three LPC quantizers. We can see

Bit error rate (in %)	Segmental SNR (in dB) with		
	24 bits/frame split vector quantizer	34 bits/frame LSF scalar quantizer	34 bits/frame LAR scalar quantizer
0.0	10.3	10.1	10.2
0.001	10.3	10.1	10.2
0.01	10.3	10.1	10.2
0.05	10.2	10.0	10.1
0.1	10.2	10.0	10.1
0.5	10.0	9.6	9.7
1.0	9.7	9.3	9.3
10.0	7.1	5.0	5.5

Table 6. Effect of channel errors on the performance (measured in terms of segmental signal-to-noise ratio (SNR) of the CELP-coded speech) of the 24 bits/frame split vector quantizer, the 34 bits/frame LSF-based scalar quantizer and the 34 bits/frame log-area ratio (LAR) based scalar quantizer.

from this table that all the three LPC quantizers show almost no degradation in the segmental signal-to-noise ratio for bit error rates up to 0.1%. For higher bit error rates, the 24 bits/frame split vector quantizer results in better signal-to-noise ratio than the 34 bits/frame scalar quantizers. Informal listening of the coded speech shows that effect of channel errors is negligible for bit error rates up to 0.1%. For higher bit error rates, the CELP-coded speech from the 24 bits/frame split vector quantizer sounds at least as

[2]In the CELP coder, used here, we do the LPC analysis every 20 ms and perform the codebook search every 5 ms. The fixed codebook index and gain are quantized using 8 bits and 5 bits, respectively. The adaptive codebook index and gain are quantized using 7 bits and 4 bits, respectively.

good as that from the 34 bits/frame scalar quantizers. Thus, we can conclude that the 24 bits/frame split vector quantizer performs at least as well as the 34 bits/frame scalar quantizers in the presence of channel errors.

Next, we study the use of error correcting codes for improving the performance of the 24 bits/frame split vector quantizer in the presence of channel errors. As mentioned earlier, the naturally-ordered codevectors in the codebook (obtained by using the splitting method for the initialization of the vector quantizer design algorithm) have the property that the most significant bits of their binary addresses are more sensitive to channel errors than the least significant bits. We use this property to our advantage by protecting the most significant bits using error correcting codes. We use here only simple error correcting codes (such as Hamming codes [14]) for protecting these bits. An (n,m) Hamming code is a block code which has m information bits and uses an additional (n-m) bits for error correction. The number of errors this code can correct depends on the values of n and m. The following two Hamming codes are investigated here: 1) (7,4) Hamming code and 2) (15,11) Hamming code. Both these codes can correct only one error occurring in any of the information bits. Recall that in the 24 bits/frame split vector quantizer, we divide the LSF vector into two parts and quantize these parts independently using two 12 bits/frame vector quantizers. We protect the most significant bits of these two vector quantizers separately. Thus, when we use the (7,4) Hamming code to protect 4 most significant bits from each of the two parts, it means that we are using an additional 6 bits/frame for error correction. Similarly, use of the (15,11) Hamming code (for protecting 11 most significant bits from each of the two parts) amounts to an additional 8 bits/frame for error correction. Performance (in terms of spectral distortion) of the 24 bits/frame split vector quantizer with these error correcting codes is shown in Tables 7 and 8, respectively, for different bit error rates. By comparing these tables with Table 3, we see that the use of error correcting

Bit error rate (in %)	Av. SD (in dB)	Outliers (in %)	
		2-4 dB	>4 dB
0.0	1.03	1.03	0.00
0.001	1.03	1.03	0.01
0.01	1.03	1.06	0.01
0.05	1.03	1.29	0.05
0.1	1.05	1.78	0.09
0.5	1.13	4.56	0.60
1.0	1.25	8.14	1.49
10.0	3.07	40.21	25.79

Table 7. Effect of channel errors on the spectral distortion (SD) performance of the 24 bits/frame split vector quantizer using 6 bits/frame for error correction.

codes improves the performance of the split vector quantizer in the presence of channel errors. In particular, when 8 bits/frame are used for error correction, we see from Table 8 that there is no degradation in performance due to the channel errors for bit error

Bit error rate (in %)	Av. SD (in dB)	Outliers (in %)	
		2-4 dB	>4 dB
0.0	1.03	1.03	0.00
0.001	1.03	1.03	0.00
0.01	1.03	1.03	0.00
0.05	1.03	1.03	0.00
0.1	1.03	1.03	0.00
0.5	1.04	1.18	0.16
1.0	1.06	1.39	0.50
10.0	3.11	17.39	31.23

Table 8. Effect of channel errors on the spectral distortion (SD) performance of the 24 bits/frame split vector quantizer using 8 bits/frame for error correction.

rates as high as 0.1%. In other words, the split vector quantizer provides transparent quantization of LPC parameters for channel error rates up to 0.1%. Also, for a bit error rate of 1%, there is very little additional distortion i.e., the average spectral distortion is still about 1 dB and outliers are few in number. Thus, the performance of the 24 bits/frame split vector quantizer using an additional 8 bits/frame for error correction is very good up to bit error rates of 1%. Similar observations can be made from Table 9, where the performance of the 24 bits/frame split vector quantizer is measured in terms of segmental signal-to-noise ratio of the CELP-coded speech. Thus, by using an

Bit error rate (in %)	Segmental SNR (in dB) using		
	0 bit/frame for error correction	6 bits/frame for error correction	8 bits/frame for error correction
0.0	10.3	10.3	10.3
0.001	10.3	10.3	10.3
0.01	10.3	10.3	10.3
0.05	10.2	10.3	10.3
0.1	10.2	10.2	10.3
0.5	10.0	10.2	10.3
1.0	9.7	10.1	10.2
10.0	7.1	8.4	8.3

Table 9. Effect of channel errors on the performance (measured in terms of segmental signal-to-noise ratio (SNR) of the CELP-coded speech) of the 24 bits/frame split vector quantizer using error correcting codes.

additional 8 bits/frame for error correction, the 24 bits/frame split vector quantizer can perform quite well over a wide range of bit error rates.

CONCLUSIONS

In this paper, we have described a split vector quantizer which requires only 24 bits/frame to achieve transparent quantization of LPC information i.e., with an average spectral distortion of about 1 dB, less than 2% outliers in the range 2-4 dB, and no outlier having spectral distortion greater than 4 dB. We have studied the effect of channel errors on the performance of this quantizer. It has been found that the split vector quantizer which employed the naturally-ordered codebooks obtained by using the splitting method for the initialization of the vector quantizer design algorithm is as robust to channel errors as the scalar quantizers.

REFERENCES

[1] P. Kroon and B.S. Atal, "Predictive coding of speech using analysis-by-synthesis techniques," in *Advances in Speech Signal Processing*, S. Furui and M.M. Sondhi, Eds. New York, NY: Marcel Dekker, 1991, pp. 141-164.

[2] B.S. Atal, R.V Cox and P. Kroon, "Spectral quantization and interpolation for CELP coders," *Proc. IEEE Int. Conf. Acoust., Speech, Signal Processing*, Glasgow, Scotland, pp. 69-72, May 1989.

[3] K.K. Paliwal and B.S. Atal, "Efficient vector quantization of LPC parameters at 24 bits/frame," *Proc. IEEE Int. Conf. Acoust., Speech, Signal Processing*, Toronto, Canada, pp. 661-664, May 1991.

[4] B. Bhattacharya, W. P. LeBlanc, S. A. Mahmoud, and V. Cuperman, "Tree searched multi-stage vector quantization of LPC parameters for 4 kb/s speech coding," *Proc. IEEE Int. Conf. Acoust., Speech, Signal Processing*, pp. 105-108, May 1992.

[5] K.K. Paliwal, "A perception-based LSP distance measure for speech recognition," *J. Acoust. Soc. Am.*, vol. 84, pp. S14-15, Nov. 1988.

[6] B.S. Atal, "Predictive coding of speech at low bit rates," *IEEE Trans. Commun.*, vol. COM-30, pp. 600-614, Apr. 1982.

[7] S. Singhal and B.S. Atal, "Improving performance of multi-pulse LPC coders at low bit rates," *Proc. IEEE Int. Conf. Acoust., Speech, Signal Processing*, San Diego, pp. 1.3.1-1.3.4, Mar. 1984.

[8] Y. Linde, A. Buzo and R.M. Gray, "An algorithm for vector quantizer design," *IEEE Trans. Commun.*, vol. COM-28, pp. 84-95, Jan. 1980.

[9] F.K. Soong and B.H. Juang, "Optimal quantization of LSP parameters," *Proc. IEEE Int. Conf. Acoust., Speech, Signal Processing*, New York, pp. 394-397, Apr. 1988.

[10] J.P. Campbell, Jr., V.C. Welch and T.E. Tremain, "An expandable error-protected 4800 bps CELP coder (U.S. federal standard 4800 bps voice coder)," *Proc. IEEE Int. Conf. Acoust., Speech, Signal Processing*, Glasgow, Scotland, pp. 735-738, May 1989.

[11] J.R.B. De Marca and N.S. Jayant, "An algorithm for assigning binary indices to the codevectors of a multidimensional quantizer," *Proc. IEEE Int. Comm. Conf.*, Seattle, pp. 1128-1132, June 1987.

[12] N. Farvardin, A study of vector quantization for noisy channels," *IEEE Trans. Inform. Theory*, vol. 36, pp. 799-809, July 1990.

[13] K. Zeger and A. Gersho, "Pseudo-Gray coding," *IEEE Trans. Commun.*, vol. 38, pp. 2147-2158, Dec. 1990.

[14] A.M. Michelson and A.H. Levesque, *Error-Control Techniques for Digital Communication.* New York, NY: John Wiley, 1985.

25

ERROR CONTROL AND INDEX ASSIGNMENT FOR SPEECH CODECS

Neil B. Cox
MPR Teltech Ltd.
8999 Nelson Way, Burnaby, B.C., Canada

This chapter describes a generalization of the pseudo-Gray coding method [2] of index assignment optimization for vector quantization codebooks. Such optimizations are an attractive means of providing error control for vector quantizers, as improved robustness to channel errors can be obtained without the addition of extra bits. The generalized optimization accounts for non-binary-symmetric channels (non-BSCs) and for interaction between index assignment and externally-applied error control. Evaluation results indicated that performance gains can be made when the assumptions of previous algorithms are violated.

THE GENERALIZED PSEUDO-GRAY CODING METHOD

The following description of vector quantization serves to fix notation. One starts by constructing a codebook (or table) of codevectors ($\mathbf{w}_r$, $r=0, ..., R-1$) such that the sequence being quantized can always be adequately represented by a series of codevectors. A unique index ($i(r)$) is assigned to each codevector ($\mathbf{w}_r$) and the indices rather than the codevectors are transmitted. A copy of the codebook is also stored in the receiver so that the received index (j) can be used to identify the most probable input codevector ($\mathbf{w}_{n(j)}$). Here the received index is converted to a vector number through reference to $n(j)$, the inverse of $i(r)$.

Pseudo-Gray coding endeavors to identify an index allocation ($i(r)$, $n(j)$) which minimizes the effect of bit errors. The method minimizes the expected value of the distance between the received codevector ($\mathbf{w}_{n(j)}$) and the codevector that would have been received on an error-free channel ($\mathbf{w}_r$). The value to be minimized is:[1,2]

$$\Lambda = E[\,d(\mathbf{w}_r, \mathbf{w}_{n(j)})\,] = \sum_{m=1}^{M} \varepsilon^m (1-\varepsilon)^{b-m} \sum_{r=0}^{R-1} C_m(\mathbf{w}_r)$$

where b is the number of bits in an index, ε is the bit-error probability for the assumed memoryless BSC, R is the number of codevectors and M is the maximum number of bits in error to be considered in the optimization ($1 \le M \le b$). $C_m(\mathbf{w}_r)$ is the average cost of an m-bit error in the index for $\mathbf{w}_r$, and is expressed by:

$$C_m(\mathbf{w}_r) = p[\mathbf{w}_r] \sum_{j \in S_m(i(r))} d(\mathbf{w}_r, \mathbf{w}_{n(j)})$$

where $p[\mathbf{w}_r]$ is the probability of $\mathbf{w}_r$, $S_m(i(r))$ is the set of all indices with a Hamming distance of m from the index for $\mathbf{w}_r$, and $d(\mathbf{w}_r, \mathbf{w}_{n(j)})$ is a suitable measure of distance between $\mathbf{w}_r$ and $\mathbf{w}_{n(j)}$.

The generalized algorithm is a natural extension of the above formulation. The assumption of a BSC channel was removed by straightforward substitution of a more general probability table, and the cost measure was modified to include the effects of forward error control (FEC). The resulting criterion is:

$$\Lambda' = E[\, d(\mathbf{w}_r, \mathbf{w}_{n(j)})\,] = \sum_{m=1}^{M} p\_err_m \, (1-\alpha_m \beta_m) \sum_{r=0}^{R-1} C'_m(\mathbf{w}_r)$$

where p_err_m is the probability of a given m-bit error pattern under the assumption that all such patterns are equi-probable for a given m, α_m is the probability of external detection of an m-bit error, and β_m is the relative benefit provided by α_m ($\beta_m = 1$ implies all detectable m-bit errors are correctable, $\beta_m = 0$ implies detection provides no benefit). The new cost measure is:

$$C'_m(\mathbf{w}_r) = p[\mathbf{w}_r] \sum_{j \in S_m(i(r))} d(\mathbf{w}_r, \mathbf{w}_{n(z(j,i(r)))})$$

where $z(j, i(r))$ = the output index produced by a FEC when $i(r)$ is the proper index but j is received.

Certain limitations should be noted when using α_m, β_m or $z(j, i(r))$ to represent a FEC. For α_m and β_m it is assumed that the benefits can be averaged across all error patterns. The effects of the FEC on undetectable error patterns are not represented, and both α_m and β_m are assumed to be independent of the index assignment. The error control represented by $z(j, i(r))$, on the other hand, simulates a relatively short block code applied on an index-by-index basis. Even this limited scenario is only true if all bits of the code are included as part of the index assignment. Nonetheless, a reasonable approximation of the effect of an FEC should be possible by setting these parameters based on a probabilistic understanding of the effect of a FEC.

Allocation of Unused Indices

A procedure is presented here to provide an intuitive means of allocating unused indices when the codebook is not fully populated. The task is to identify an index allocation ($i(r), n(z(j, i(r)))$) and an error control mapping ($z(j, i(r))$) that minimize Λ'. The natural assumption when $j = i(r)$ for some r is to set $z(j, i(r)) = j$. The problem then becomes one of optimizing the index assignment $i(r)$ and its inverse $n(j)$, with special measures taken for the extra entries in $n(j)$. The following procedure is proposed: 1) Provide an initial specification for $i(r)$ and set the corresponding entries in the inverse function $n(j)$. The remaining entries in $n(j)$ represent detectable errors. 2) Optimize the index assignment under the assumption that the distance is zero when a detectable error is encountered. This assigns the unused indices to potentially beneficial positions. 3) Connect each unused index to the codevector that produces the smallest increase in distortion relative to the zero distance assumption stated above. The unused indices are set one at a time. 4) Repeat the optimization of step 2 with the zero distance assumption removed.

EVALUATIONS

Evaluations were performed using the residual vector codebook of a CELP-class codec. This evaluation included tests of the relative benefit of generalized pseudo-Gray coding for trained and untrained codebooks, tests of the incremental benefit provided by redundant indices, and tests of the effectiveness when applied in tandem with simulations of externally-applied error control. Two codebooks were used. The first codebook (the Gaussian codebook) contained 128 random Gaussian vectors, each comprised of 8 elements. The second codebook (the trained codebook) was derived using the LBG algorithm initialized with the first codebook. All optimized index assignments for the Gaussian codebook were obtained under the assumption that vectors are equi-probable. Except for cases where comparisons were made with the Gaussian codebook, the vector probabilities for the trained codebook were set according to the frequency-of-use statistics generated during the training process.

The measure of distortion for a given index assignment and channel simulation was the average Euclidean distance between the desired codevector and the codevector that is actually selected based on the received and possibly corrupted index. This was normalized with respect to the expected distance for a random received index, i.e., for a BSC with BER=0.5. Thus:

$$DISTORTION\ (dB) = 20\log_{10}(E[\,d(\mathbf{w}_{\pi(i)}, \mathbf{w}_{\pi(j)})\,] \,/\, E[\,d(\mathbf{w}_{\pi(i)}, \mathbf{w}_x)\,])$$

where $\mathbf{w}_x$ is a randomly chosen codevector, i is the transmitted index and j is the received index. This metric must be a large negative number for acceptable communication, as a value of 0 dB implies that the received index is no better than a randomly-chosen index.

The evaluations entailed deriving the worst-case and the best-case index assignments for each of the codebooks under a range of conditions. A Euclidean distance was used in all cases to measure the dissimilarity between vectors. The local maxima or minima in distortion were found using a modification of the the binary switching algorithm described by Chen and Gersho.[1] The modified algorithm reassigns the index with the highest cost rather than reassigning the index for the codevector with the highest cost. That is, the procedure now starts by finding the index that has the highest cost, and then reduces the distortion, if possible, by swapping it with another index. This is functionally equivalent to the old procedure for fully populated codebooks. However, the modification is needed when redundant indices are present to ensure that all possible swaps are considered.

RESULTS

Some results of applying the generalized pseudo-Gray coding method for a memoryless BSC are illustrated in Figure 1. Data are for 7-bit indices assigned to the trained codebook. It is apparent that the distortion at a given BER varied by about 4 dB, depending on the index assignment. Figure 1 also indicates that the use of a faulty vector probability assumption can be significant, with a cost of about 0.5 dB when evenly-distributed vector probabilities were substituted. The results were substantially the same for the Gaussian codebook.

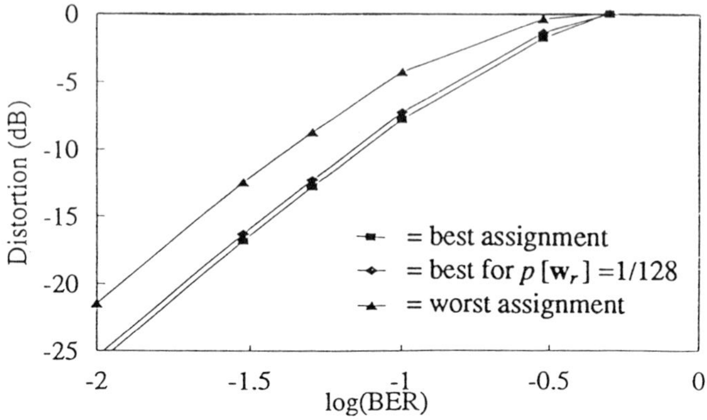

Figure 1: Distortion vs Bit-Error Rate for Optimized Index Assignments.

Figure 2 illustrates the effect of protecting some of the bits of the indices by external error control. The analysis conditions were the same as for Figure 1 except that the bit-error rate was fixed at 0.01. The protection was simulated by constraining $S_m(i(r))$ such that certain bits were error-free. The distortion initially improved by about 2 dB per protected bit, with larger gains obtained when the majority of index bits were protected. Reoptimization of the index assignment provided a further gain of about 0.5 dB when a minority of the bits were protected, and a further gain that approached 3.3 dB when most of the bits were protected. In addition, reoptimization provided about a 1 dB gain for the single-bit error correction scenario represented by setting $\alpha_1\beta_1 = 1$.

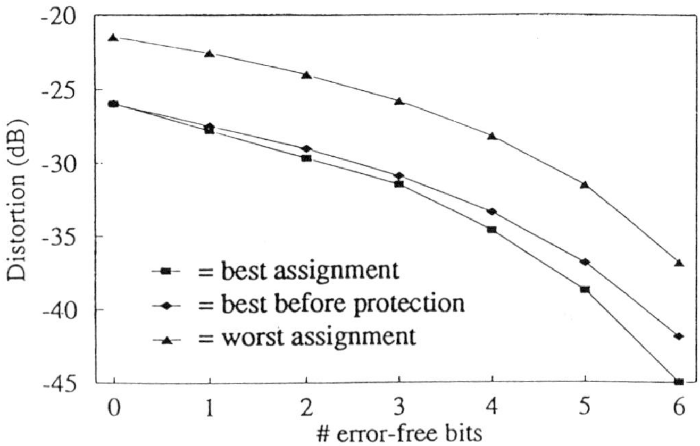

Figure 2: Effect of External Protection of Index Bits on Optimized Index Assignments (BER=0.01).

Figure 3 demonstrates the utility of the unused-index-allocation strategy. The analysis conditions were the same as for Figure 2. The trained codebook was shortened one vector at a time by replacing the two "most similar" vectors with a probability-weighted mean vector. A probability-weighted Euclidean distance was used as a measure of similarity, and the probability of the derived replacement vector was set equal to the sum of the two input probabilities. The distortion for optimized indices steadily decreased as vectors were removed, culminating in a 2 dB improvement when the codebook size was halved. This corresponded well with results obtained when an eighth index bit was included: the unused-index-allocation procedure provided a 2 dB improvement over the case where the new indices were forced to be a replication of the best 7-bit index assignment.

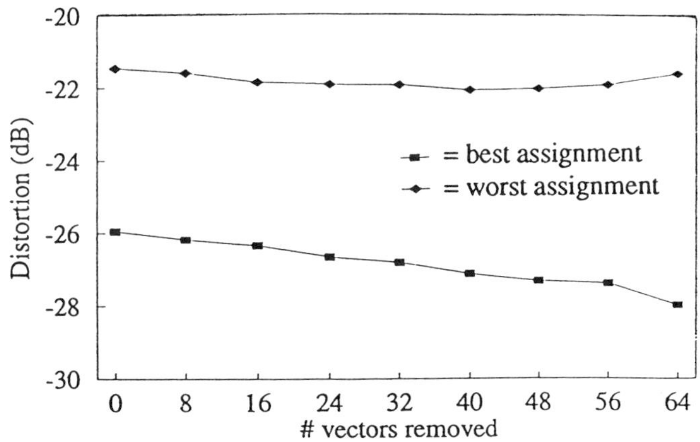

Figure 3: Effect of Redundancy Allocation after Vector Removal on Optimized Index Assignments (BER=0.01).

In conclusion, the generalized pseudo-Gray algorithm for index assignment optimization combined with the allocation strategy for unused indices was shown to provide modest gains when assumptions for the original algorithm were violated. Examples include a 0.5 dB improvement when a few of the index bits were externally protected, a 1 dB improvement when single-bit error correction was simulated, and a 2 dB improvement when an extra index bit was added. It is worth noting that it was sometimes necessary to use $M > 1$ to fully obtain these gains. The daunting computational burden of this can be minimized by using $M = 1$ in a preliminary optimization, and then progressively incrementing it until no improvement is derived. It was generally sufficient to stop at $M = 2$.

REFERENCES

[1] Chen, J.H., Davidson, G., Gersho, A., and Zeger, K., ''Speech Coding for the Mobile Satellite Experiment, '' *IEEE Int. Conf. on Commun.*, 1987, pp. 756-763.

[2] Zeger, K. and Gersho, A., ''Pseudo-Gray Coding, '' *IEEE Trans. on Commun.*, 1990, pp. 2147-2158.

PART VII

TOPICS IN SPEECH CODING

This section is dedicated to new techniques that improve the performance of existing speech coding systems. The subjects covered include the design of the long-term predictor (adaptive codebook) and of the excitation codebooks in CELP, LPC parameters quantization, improvements of the excitation in LPC vocoders. The chapters by Gerson and Jasiuk and Veeneman and Mazor are dedicated to efficient techniques for determining the parameters of the long-term (pitch) predictors in CELP environment. New techniques for designing the excitation codebooks in CELP are presented in chapters by Taniguchi et al, Dymarksi and Moreau, and Benyassine et al.

An efficient representation of the CELP excitation using non-uniform frequency-domain sampling is presented by Gupta and Atal. Wang et al present new results on the LPC parameters quantization using a general product vector quantization approach. McCree et al present an improved 2.4 kb/s LPC vocoder with frequency-dependent mixed excitation. Finally, Bhaskar presents a hybrid system using adaptive predictive coding and transform domain quantization.

26

EFFICIENT TECHNIQUES FOR DETERMINING AND ENCODING THE LONG TERM PREDICTOR LAGS FOR ANALYSIS-BY-SYNTHESIS SPEECH CODERS

Ira A. Gerson and Mark A. Jasiuk

Corporate Systems Research Laboratories
Motorola
1301 E. Algonquin Road, Schaumburg, IL 60196

INTRODUCTION

Many analysis-by-synthesis speech coders, such as CELP coders, make use of a combination of long-term and short-term predictors. The use of long term predictors (adaptive codebooks) incorporating lags with sub-sample resolution has contributed to enhanced performance for these coders, particularly for high pitched speakers [1],[2]. This paper discusses an efficient technique for determining the lag for the long-term predictor (index of the adaptive codebook) when sub-sample resolution lags are allowed. Also, an efficient technique for encoding these lags (adaptive codebook indices) is presented.

In general a full search of the adaptive codebook with sub-sample lag resolution results in a substantial increase in the coder's computational requirements over a coder incorporating an adaptive codebook limited to integer lags. An efficient lag search algorithm combining open-loop and closed-loop processing is described in the context of independent coding of each lag.

During voiced speech, the long term predictor (LTP) lags exhibit a high degree of correlation from subframe to subframe; a fact which is not exploited when the lags are coded independently. A number of methods have been proposed which exploit this correlation to code the LTP lags. One technique codes the frame lag and the LTP lag deviations relative to the frame lag at each subframe [3],[4]. This method, however, does not yield maximum coding efficiency; a deviation at each subframe needs to be specified in addition to the frame lag. In [5] the lag is coded independently at odd subframes and delta coded at even subframes. The independently selected lag determines the search bounds for the lag in the following subframe. This can result in suboptimal lag coding since odd subframe lags are coded without considering the impact of that coding on the next (even) subframe, which may degrade performance, especially in transition regions. The methodology for the efficient LTP lag search, described here, is extended to a trajectory based lag coding scheme, where a *frame lag trajectory* is defined to be a sequence of subframe lags within the frame. The first subframe's lag is coded independently, with each subsequent subframe's lag delta coded relative to the preceding subframe's coded value of the lag. The frame lag trajectory is globally optimized, open-loop, over all subframes in the frame and allows for a closed-loop lag search at each subframe to refine the lag estimate.

EFFICIENT LONG TERM ADAPTIVE CODEBOOK SEARCH

Full search of the adaptive codebook results in significantly higher complexity when sub-sample resolution lags are allowed. To retain the performance advantage due to high resolution lags while keeping complexity in check, a two stage hybrid open-loop/closed-loop search may be used for the adaptive codebook. This approach is similar to the hybrid open-loop/closed-loop search of Chen et al. [3] and the restrictive pitch deviation coding technique of Yong and Gersho [4]. The open-loop stage determines a list of candidate lags to be evaluated in the closed-loop search. Let $C_o(k)$ be the correlation corresponding to integer lag k, in the open-loop sense:

$$C_o(k) = \sum_{n=0}^{N-1} w(n)\, w(n-k), \qquad \text{for } k=L_{min}, \ldots, L_{max} \tag{1}$$

and define $G_o(k)$ as:

$$G_o(k) = \sum_{n=0}^{N-1} w^2(n-k), \qquad \text{for } k=L_{min}, \ldots, L_{max} \tag{2}$$

where w(n) is the spectrally weighted input speech, N is the number of samples in a subframe and L_{min} and L_{max} specify the range of integer lags. The spectrally weighted input speech is used so that the open loop search uses a selection criterion which is similar to that used by the closed loop search which is based on the weighted speech signal. J, the lag which maximizes the prediction gain of a first order integer lag pitch predictor over the weighted speech for the subframe, can be found by setting J to the value of k which maximizes the *normalized correlation function*:

$$\frac{C_o(k)}{\sqrt{G_o(k)}}, \qquad k = L_{min}, \ldots, L_{max} \tag{3}$$

Once the best integer open-loop lag, J, has been obtained, submultiples of J are evaluated to see if they are local maxima of the normalized correlation function. *Allowable lags* are defined as those lag values, both integer and fractional, which can be represented by the lag quantizer. If an integer resolution local maximum is found, the C_o and G_o arrays are interpolated around this integer lag to find the sub-sample resolution maximum in the normalized correlation function which corresponds to an allowable lag. The estimated prediction gain due to the sub-sample resolution maximum is then compared to the prediction gain computed for lag J. If it exceeds a specified percentage of that gain, it is classified as a *surviving peak*. The lowest lag value corresponding to a surviving peak is the *minimum lag surviving peak*. The multiples of the minimum lag surviving peak are then evaluated in a similar fashion. The output of this process is a list of sub-sample resolution lags which is reordered according to prediction gain.

The closed-loop adaptive codebook search is based on the list of lags from the open-loop search. The closed-loop search evaluates a range of allowable lags around each of the top few surviving peaks for each subframe.

There are several advantages of this hybrid open-loop/closed-loop adaptive codebook search procedure. An intelligent choice is made in determining a subset of lags to be searched closed-loop. This limits the amount of computation. For voiced subframes, where the adaptive codebook vector dominates the excitation, there is a high degree of correlation between the estimated open-loop peaks and the lag selected in an exhaustive closed-loop search. Multiple peaks are allowed to be searched. The ordering of peaks based on prediction gain, is designed to maximize the coder performance when the number of peaks to be searched is constrained. Also, complexity scaling is easily achieved by appropriately selecting the number of surviving peaks and the number of allowable lags to be evaluated in the closed-loop search. For unvoiced subframes there is less similarity between the open-loop and closed-loop long term correlations, but the adaptive codebook vector contribution to the excitation is less important in this case.

Table 1 shows the performance of a 6.9 kb/s VSELP speech coder incorporating sub-sample resolution lags and harmonic noise weighting (HNW) [6],[7] for three different methods of lag search. The results are given in terms of the spectrally and harmonically weighted error over a ninety second speech database for each method. The hybrid method utilizes at most two surviving open-loop peaks, and evaluates three allowable lags for each of the two peaks. If there is only one surviving peak, five allowable lags are evaluated. Therefore at most six lags are evaluated closed-loop. Even with just six closed-loop lag evaluations, the hybrid lag search performs almost as well as an exhaustive closed-loop search. Removing the harmonic noise weighting from the lag search does not affect performance.

LAG SEARCH METHOD	$WSNR_{seg}$ (dB)	$WSNR_{total}$ (dB)
full search with HNW	12.47	18.46
hybrid search with HNW	12.22	18.28
hybrid search, no HNW	12.24	18.31

Table 1 – LAG Search Performance

FRAME LAG TRAJECTORY DERIVATION

The efficient lag search technique is now extended to frame trajectory based lag encoding. A *frame lag trajectory* is defined to be a sequence of subframe lags within a frame. Given N_s subframes per frame, the first subframe's lag is coded independently, with each subsequent subframe's lag being delta coded relative to the preceding subframe's coded lag value. One weakness of the delta lag encoding method, as it is usually implemented, stems not from the coding method itself, but from the sequential selection process of the subframe lags. This may result in a suboptimal frame lag trajectory, thus degrading the LTP performance over the frame. The method attempts to globally optimize the frame lag trajectory over the whole frame. Although the frame lag trajectory is derived open-loop, it allows for closed-loop refinement of the lag within M_C allowable lag values relative to the open-loop lag value at each subframe. This assures that any combination of the lags selected closed-loop satisfies the delta coding constraints.

The method assigns F bits to code the first subframe's lag and D bits to code each of the (N_S-1) delta lags , defining $2^{[F+(N_S-1)D]}$ possible lag trajectories per frame. The delta coding can code lags within $-2^{(D-1)}$ to $2^{(D-1)}-1$ allowable lag levels of the previous subframe's coded lag value. For reasonable values of F, D, and N_S, evaluation of all trajectories at a frame is impractical. Instead a small subset of the frame lag trajectories is evaluated, from which the trajectory yielding the highest open-loop LTP frame prediction gain is selected.

The process for obtaining a list of lags corresponding to the maxima in $\frac{C_o(k)}{\sqrt{G_o(k)}}$ at a given subframe has already been described. The lags in the list are ordered according to prediction gain. Assume that such a list is generated for each subframe, and that the C_O and G_O arrays for each subframe are also available. The top few lags are selected from the list of lags at each subframe to become *anchor lags* for potential frame lag trajectories. For each anchor lag, a frame lag trajectory is constructed using the anchor lag and its associated subframe as the starting point. The trajectory is extended in the forward direction to the last subframe of the frame and in the backward direction to the first subframe of the frame. When extending the trajectory in the forward direction, the lag for the next subframe must be within $-2^{(D-1)}+M_C$ to $2^{(D-1)}-1-M_C$ allowable lag levels of the current subframe's lag. The lag which maximizes $\frac{C_o(k)}{\sqrt{G_o(k)}}$ within the allowable range is selected as the next subframe's lag for the current trajectory. When extending the trajectory in the backward direction, the lag for the previous subframe must be within $-2^{(D-1)}+1+M_C$ to $2^{(D-1)}-M_C$ allowable lags of the current subframe's lag.

Each frame lag trajectory which has been evaluated at the current frame is stored. If an anchor lag under consideration is already part of a previously evaluated frame lag trajectory, a new frame lag trajectory will not be evaluated for that anchor lag. Instead, the next lag from the list of lags at that subframe which is not part of a previously evaluated frame lag trajectory, becomes the new anchor lag. If the list of lags at that subframe does not contain such a replacement candidate, the evaluation of trajectories anchored at that subframe ends. Since each subframe has associated with it a set of anchor lags to be evaluated, the choice of initial subframe for anchoring the potential frame lag trajectories is not critical. Thus a set of possible frame lag trajectories is derived.

The trajectory with the highest open-loop prediction gain for the frame is selected from the set. Note that the open-loop search range for delta coding is reduced by M_C levels at each extreme of the range to allow for closed-loop evaluations of

$2M_C+1$ allowable lag values per subframe around the open-loop lag defined by the selected trajectory. This ensures that any combination of the lags selected closed-loop may be delta coded with $F+(N_S-1)D$ bits per frame.

Table 2 compares the performance a VSELP speech coder using three different techniques for coding the lags. The first technique uses frame lag trajectory (FLT) based LTP encoding. The second technique delta codes the lags without frame lag trajectory optimization and the third technique independently codes the LTP lags (8 bits/subframe). In both delta coded cases, 8 bits are allocated for independently coding the first subframe's lag and 4 bits/subframe specify the lag delta codes for the remaining three subframes of the frame. A hybrid LTP lag search, with no HNW, is employed in each case, with M_C set to 1. For the independently coded LTP lags, the hybrid open-loop/closed-loop lag search algorithm is used, but with the closed-loop lag search restricted to vicinity of the best open-loop lag at a given subframe. Up to two anchor lags/subframe are allowed for the FLT based LTP encoding. In the delta coding scheme without frame lag trajectory optimization, the lag found closed-loop in the vicinity of the allowable lag corresponding to the best open-loop correlation peak at the first subframe, anchors the frame lag trajectory. The results have been obtained over a ninety second speech database and are expressed in terms of the spectrally and harmonically weighted error. This speech database is different from the database used for Table 1, so the results in Table 1 and Table 2 may not be directly compared. The ranking is as expected, with the independently coded LTP lags performing best, the optimized FLT placing second, and delta coding of LTP lag without FLT optimization placing third. What the numbers do not emphasize is that perceptually, the first two systems are very close. The optimization of the frame lag trajectory effectively eliminates the artifacts which the delta coding scheme without FLT occasionally introduces.

LAG SEARCH METHOD	$WSNR_{seg}$ (dB)	$WSNR_{total}$ (dB)
lag coded independently	13.81	18.35
delta coded lag, FLT	13.67	18.19
delta coded lag, no FLT	13.46	17.70

Table 2 – LAG Search Performance

The output of the trajectory search is a list of lags to be evaluated closed-loop at each subframe, and the open-loop LTP prediction gain for the selected frame lag trajectory. The high degree of subframe to subframe correlation among the lags, evident for voiced speech frames and efficiently exploited by the delta coding scheme described, is not present in unvoiced speech frames. Consequently, the delta coding of the lags can degrade the coder's performance for unvoiced speech. To improve coder performance for unvoiced speech, the long term predictor may be deactivated and the LTP bits reallocated to an additional codebook excitation. The open-loop LTP prediction gain due to the frame lag trajectory may be used as a criterion to select between an adaptive codebook or other codebook excitation.

CONCLUSIONS

An efficient method for determining the long term predictor lag through the use of a hybrid open/closed loop search procedure has been presented. A method for delta coding the LTP lags was described which exploits differential lag coding while eliminating the performance degradation typically incurred. The performance of the coder may be improved for unvoiced speech by disabling the adaptive codebook for unvoiced frames, and reallocating the adaptive codebook bits to additional stochastic excitation.

REFERENCES

[1] P. Kroon and B.S. Atal, "Pitch Predictors with High Temporal Resolution," *Proc. IEEE Int. Conf. on Acoustics, Speech and Signal Processing*, pp. 661-664, April 1990.

[2] J.S. Marques, I.M. Trancoso, J.M. Tribolet, and L.B. Almeida, "Improved Pitch Prediction with Fractional Delays in CELP Coding," *Proc. IEEE Int. Conf. on Acoustics, Speech and Signal Processing*, pp. 665-668, April 1990.

[3] J-H Chen, R. Danisewicz, R. Kline, D. Ng, R. Valenzuela, and B. Villella, "A Real-Time Full Duplex 16/8 KBPS CVSELP Coder with Integral Echo Canceller Implemented on a Single DSP56001," *Advances in Speech Coding*, pp. 299-308, Kluwer Academic Publishers, 1991.

[4] M. Yong and A. Gersho, "Efficient Encoding of the Long-Term Predictor in Vector Excitation Coders," *Advances in Speech Coding*, pp. 329-338, Kluwer Academic Publishers, 1991.

[5] J. Campbell, V. Welch, and T. Tremain, "An Expandable Error-Protected 4800 bps CELP Coder," *Proc. IEEE Int. Conf. on Acoustics, Speech and Signal Processing*, pp. 735-738, May 1989.

[6] I.A. Gerson and M.A. Jasiuk, "Vector Sum Excited Linear Prediction (VSELP) Speech Coding at 8 kbps," *Proc. IEEE Int. Conf. on Acoustics, Speech and Signal Processing*, pp. 461-464, April 1990.

[7] I.A. Gerson and M.A. Jasiuk, "Techniques for Improving the Performance of CELP Type Speech Coders," *Proc. IEEE Int. Conf. on Acoustics, Speech and Signal Processing*, pp. 205-208, April 1991.

27

STRUCTURED STOCHASTIC CODEBOOK AND CODEBOOK ADAPTATION FOR CELP

Tomohiko Taniguchi, Yoshinori Tanaka, Yasuji Ohta

Fujitsu Laboratories Ltd.,
1015 Kamikodanaka, Nakahara-ku
Kawasaki 211, Japan

INTRODUCTION

Since its introduction in 1984, Code Excited Linear Prediction (CELP) [1] has been intensively investigated as a promising coding algorithm for providing good quality speech at low bit rates. CELP is the name for a class of coding algorithms that employs vector quantization (VQ) using a perceptually weighted error criterion measured in an Analysis-by-Synthesis loop. This process gives an efficient representation of the excitation signal and exhibits better performance than conventional coding methods. However, the codebook search requires a huge computational load, which is a major drawback in the practical implementation of CELP. In particular, for digital cellular communications, which is considered the biggest application for low bit-rate speech coding, reducing the complexity of CELP is important for small hardware size and low power consumption.

In the last few years, several computational reduction methods have been studied [2], and some of them, using structured stochastic codebooks, have achieved a good compromise between complexity and performance [3-6]. We have already proposed a hexagonal lattice codebook [7] and a sparse-delta codebook [8] effective in reducing the complexity. As an extension of the delta codebook, we propose a tree-structured delta codebook which not only reduces the complexity but reduces the memory requirements of CELP. Also, a method for adapting the distribution of the codebook based on the input speech signal is investigated for improved CELP performance.

In this chapter, the tree-structured delta codebook is first introduced, and its effectiveness in reducing the complexity of the CELP stochastic codebook search is discussed. Next, the codebook adaptation method is described which, using the special nature of the tree-structured delta codebook, controls the distribution of code vectors adaptively based on the input speech. Finally, the performance of both the codebook adaptation method and a CELP coder that uses the tree-structured delta codebook are analyzed.

TREE-STRUCTURED DELTA CODEBOOK

Codebook Structure

The tree-structured delta codebook is a variation on the delta codebook which we proposed in [8]. In the delta codebook, the differences between consecutive code vectors are stored as a delta vector codebook, instead of storing each code vector independently. Thus, each code vector (C) of the delta codebook is generated from the previous code vector and delta vector (ΔC) recursively, according to the following expression:

$$C_k = C_{k-1} + \Delta C_k \qquad \text{: Delta codebook}$$

By designing the delta vector codebook as a sparse codebook, the complexity for the stochastic codebook search can be reduced to 1/10 of the conventional method [8]. However, since the sparse-delta codebook did not reduce the memory for codebook storage, NxM words of memory are needed to store an N-dimensional delta vector codebook of size M.

To reduce the memory requirement and the complexity, the expression for code vector generation is modified to expression (1). Code vectors generated according to this expression form a tree structure as shown in Figure 1, and so we call this codebook the "tree-structured delta codebook" (or "tree-delta codebook"). A tree-delta codebook with (2^L - 1) code vectors can be generated from only L kinds of delta vectors, including an initial vector, ΔC_0 ($=C_0$) $\sim \Delta C_{L-1}$. By adding one zero-vector to the codebook, an L-bit codebook (size: 2^L) is constructed. This means that a tree-delta codebook of size M requires only NxL words of memory, (where $L = \log_2 M$).

$$C_{2k+1} = C_k + \Delta C_i \,, \quad C_{2k+2} = C_k - \Delta C_i \qquad (1)$$
$$(i=1 \sim L-1, \; 2^{i-1} - 1 \leqq k < 2^i - 1)$$

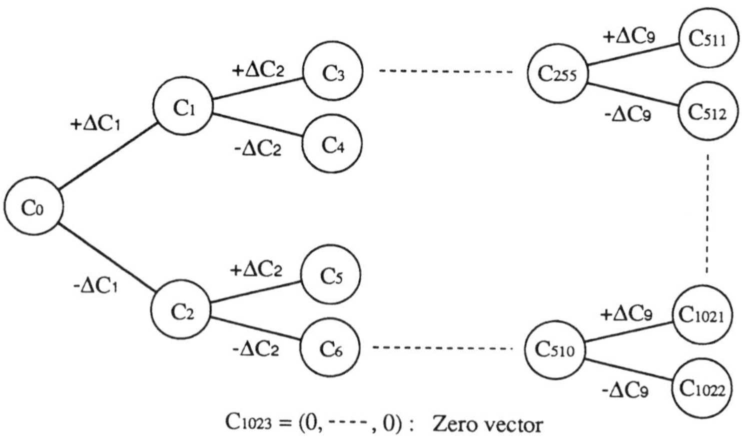

Figure 1. Tree-structured delta codebook

Reduction of the Codebook Search Complexity

The goal in a CELP stochastic codebook search is to find the code vector (C) which minimizes the error ($|E|^2$) between the input (AX) and reproduced speech (gAC). Since this process requires synthesis (C→AC) during analysis, it is called Analysis-by-Synthesis. (Here, the matrix "A" represents the weighted LPC synthesis filter 1/A'(z)). Instead of evaluating the error power, the optimal code vector can be determined equivalently by maximizing the function Rxc^2/Rcc, as in expression (2), where Rxc is the correlation between target vector (AX) and weighted code vector (AC), and Rcc is the energy of the weighted code vector (AC). As shown in Figure 2, during the stochastic codebook search, these three elements: i) filter, ii) correlation, and iii) energy, have to be calculated for each code vector. Thus, if a conventional full-gaussian codebook is used, the required numbers of calculations for a codebook search is proportional to the size of the codebook (M).

$$|E|^2 = |AX - gAC|^2 \rightarrow \min$$
$$C = \mathrm{argmax}(Rxc^2 / Rcc) \quad (2)$$
$$g = Rxc / Rcc \quad (3)$$

i) $C \rightarrow AC$ (Filter)
ii) $Rxc = (AX)^T AC$ (Correlation)
iii) $Rcc = (AC)^T AC$ (Energy)

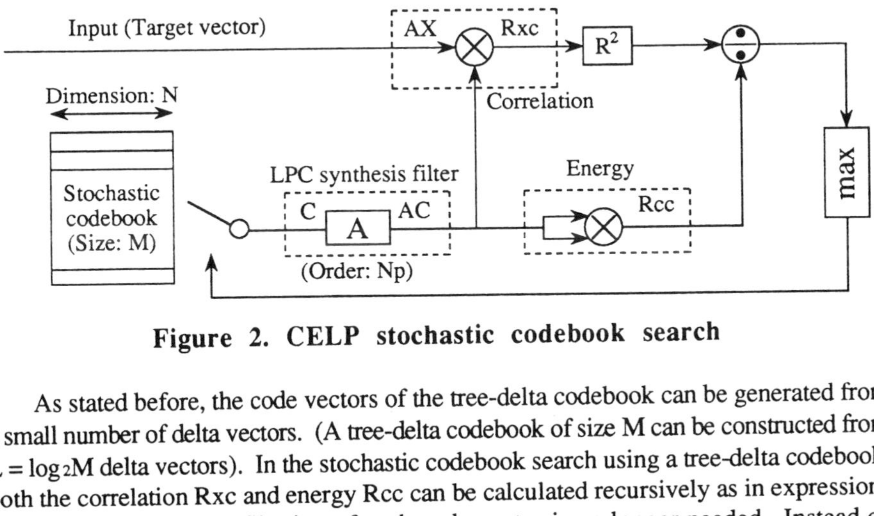

Figure 2. CELP stochastic codebook search

As stated before, the code vectors of the tree-delta codebook can be generated from a small number of delta vectors. (A tree-delta codebook of size M can be constructed from $L = \log_2 M$ delta vectors). In the stochastic codebook search using a tree-delta codebook, both the correlation Rxc and energy Rcc can be calculated recursively as in expressions (5) and (6). Therefore, filtering of each code vector is no longer needed. Instead of calculating the correlation Rxc for each code vector, the L delta vectors are first filtered, and the L correlations between target and delta vectors are calculated. For the energy term Rcc, L auto-correlations and ${}_LC_2 = L(L-1)/2$ cross-correlations are calculated among the filtered delta vectors.

$$AC_{2k+1} = AC_k + A\Delta C_i, \qquad AC_{2k+2} = AC_k - A\Delta C_i \quad (4)$$
$$Rxc^{(2k+1)} = Rxc^{(k)} + (AX)^T A\Delta C_i, \qquad Rxc^{(2k+2)} = Rxc^{(k)} - (AX)^T A\Delta C_i \quad (5)$$
$$Rcc^{(2k+1)} = Rcc^{(k)} + 2(AC_k)^T A\Delta C_i + (A\Delta C_i)^T A\Delta C_i$$
$$Rcc^{(2k+2)} = Rcc^{(k)} - 2(AC_k)^T A\Delta C_i + (A\Delta C_i)^T A\Delta C_i \quad (6)$$

The total amount of computations required for a 40-dimensional 10-bit codebook search can be reduced to 1/70 of the conventional method, as summarized in Table 2. (In this case, the complexity was estimated for a full codebook search. Further reduction can be achieved using a tree-search of the tree-delta codebook).

		Full-gaussian		Tree-delta	
Complexity	Filter	Np×N×M	400 K mac	Np×N×L	4 K mac
	Correlation	N×M	40 K mac	N×L	0.4 K mac
	Energy	N×M	40 K mac	$N\times\frac{L(L+1)}{2}$	2.2 K mac
	Total		480 K mac (96 Mops)		6.6 K mac (1.3 Mops)
Memory		N×M	40 K word	N×L	0.4 K word

mac(= multiply and accumulate) values are calculated for 40-dimensional, 10-bit codebooks

Table 1. Complexity and memory requirement

CODEBOOK ADAPTATION

Distribution of Code Vectors

Prior to discussing the performance of a simulated CELP coder, the geometric distribution of code vectors in the tree-delta codebook is observed. Although the dimension of the codebooks (N) commonly applied to CELP is 40, the 3-dimensional case (N=3) is examined for simplicity. In this example, 3 unit vectors along each axis are chosen as delta vectors, and the same vectors are used as basis vectors for a VSELP codebook [6].

As shown in Figure 3, the code vectors of the VSELP codebook are distributed uniformly in 3-D space. 6 different distributions are possible for the tree-delta codebook (only 3 are shown), each of which corresponds to one surface of the VSELP cube. For example, in case #1, the distribution of the tree-delta codebook is concentrated around the y-axis. This is because the y-directed unit vector (e_y) is set as the initial vector (C_0). If e_z is set as the initial vector, the distribution is concentrated around the z-axis (case #2).

This feature can be explained by looking at the structure of the codebook shown in Figure 1. Most of the code vectors in the tree-delta codebook contain the delta vectors in upper layers (ΔC_0 (=C_0), ΔC_1) as components. On the other hand, the delta vectors in lower layers (ΔC_{L-2}, ΔC_{L-1}) are contained in less of the code vectors. Thus, the distribution of the tree-delta codebook is concentrated on the space around the upper layer delta vectors. This implies that the distribution of the codebook depends on the order of the delta vectors, and it can be changed by rearranging the order.

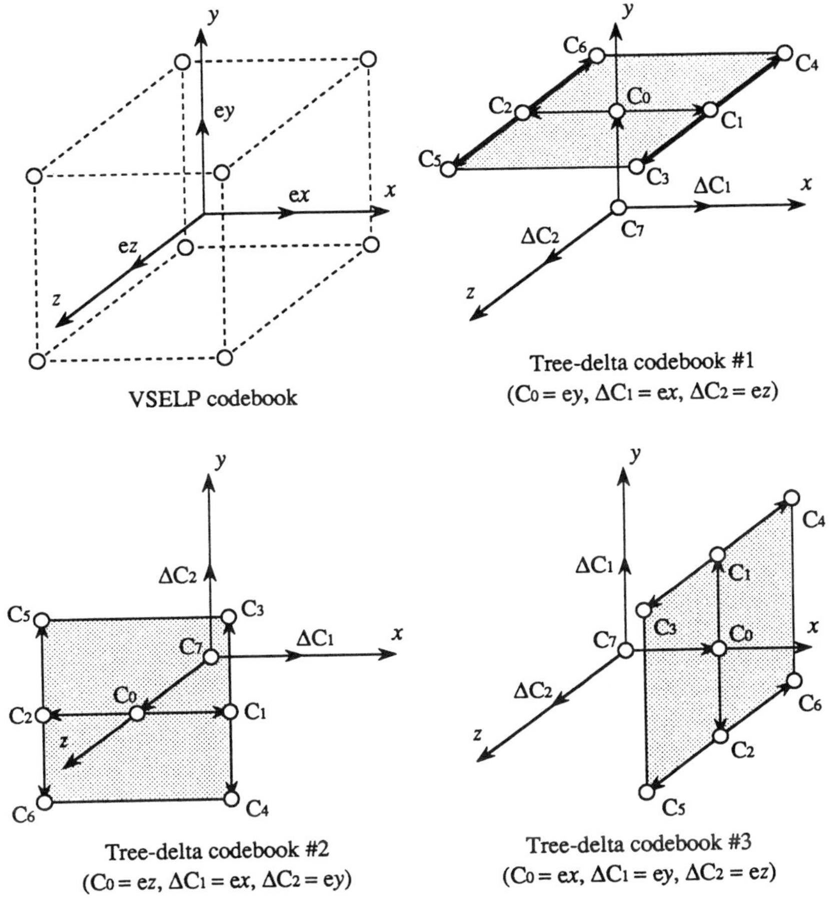

Figure 3. Distribution of code vectors

Delta Vector Sorting

The Analysis-by-Synthesis-based vector quantization carried out in CELP uses the perceptually weighted error criterion, where the characteristics of the weighted LPC synthesis filter is time-varying. Considering the special geometry of the tree-delta codebook, and applying it to the A-b-S based VQ, we propose a codebook adaptation method which controls the stochastic codebook distribution according to the synthesis filter characteristic.

As shown in Figure 4, the spherical distribution of the LPC excitation signal vectors is transformed to the elliptic distribution of the reproduced vectors by the weighted LPC synthesis filter (A). Thus, if the direction that is most amplified by the filter is known, then more code vectors can be distributed in that direction (instead of distributing code vectors uniformly), and the performance of vector quantization can be improved.

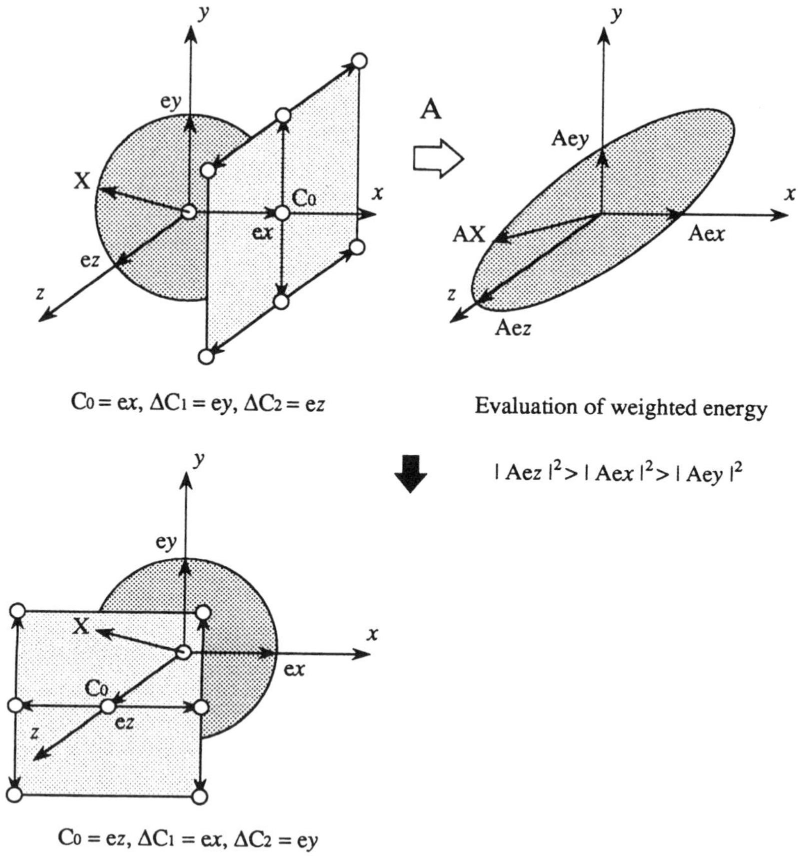

Figure 4. Delta vector sorting

The feasibility of changing the codebook distribution makes the tree-delta codebook especially suitable for this purpose. The method, which we call "delta vector sorting", controls the codebook distribution adaptively based on the synthesis filter characteristics. In this scheme, the weighted energy of each delta vector is evaluated, and the order of the delta vectors is arranged according to the amplification ratio, so that the most amplified delta vector is set to the initial vector C_0, the second most amplified one to ΔC_1, and so on (Figure 4). (For codebook adaptation using a conventional codebook, switching among codebooks designed for each filter characteristic would have to be performed. The memory for codebook storage makes this implementation impractical).

The configuration of a tree-delta fast stochastic codebook search with delta vector sorting is shown in Figure 5. In this system, the weighted energy of each delta vector is evaluated prior to the codebook search, and the order of delta vectors is arranged according to the energy. This order is determined in the same way at the decoder, so no additional information is necessary to specify the order of delta vectors.

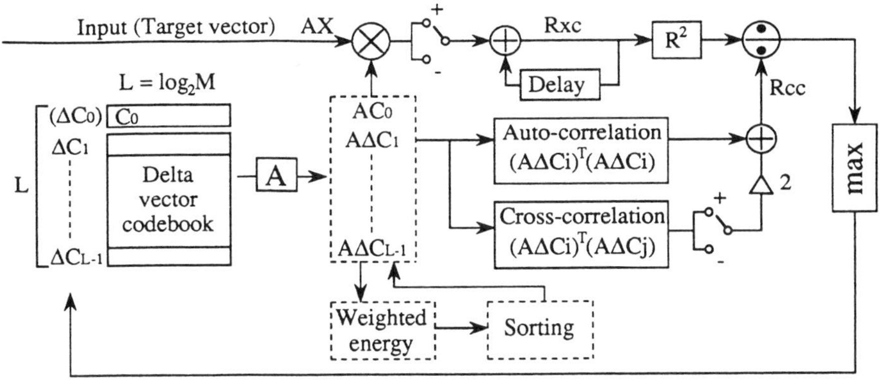

Figure 5. Tree-delta fast search with delta vector sorting

PERFORMANCE

The performance of four CELP coders with structured stochastic codebook were evaluated at 4.8 kb/s. The first one uses the VSELP codebook, while the other three use the tree-delta codebook. (10 delta/basis vectors are used to construct the 40-dimensional 10-bit codebook). Objective performance results are summarized in Table 2. The tree-delta codebook without delta vector sorting exhibits almost the same performance as the VSELP codebook, but 0.5 dB of improvement in segmental SNR was achieved by employing delta vector sorting. LPC CepstrumDistance (CD) calculated from reproduced speech was also improved 0.2 dB. To further improve the adaptation of the codebook, the expanded delta vector sorting shown in Figure 6 was applied. In this method, the 10 most amplified vectors are selected adaptively out of 40-orthonormal vector sets prepared as candidates for delta vectors. This helps increase the flexibility of the codebook distribution, and enhances the effect of delta vector sorting. The resulting SNRseg improvement was 0.9 dB, and LPC-CD was further improved. In subjective listening tests, speech reproduced by the tree-delta codebook with delta vector sorting contained less audible quantization noise, compared with the VSELP and tree-delta without delta vector sorting. The consistent improvement in perceptual quality was achieved by adopting expanded delta vector sorting.

	VSELP codebook	Tree-delta codebook		
		without delta vector sorting	with delta vector sorting	Expanded delta vector sorting
SNRseg	11.5 dB	11.6 dB	12.1 dB	12.5 dB
LPC-CD	2.6 dB	2.6 dB	2.4 dB	2.3 dB

Table 2. Simulation results for 4.8 kb/s CELP

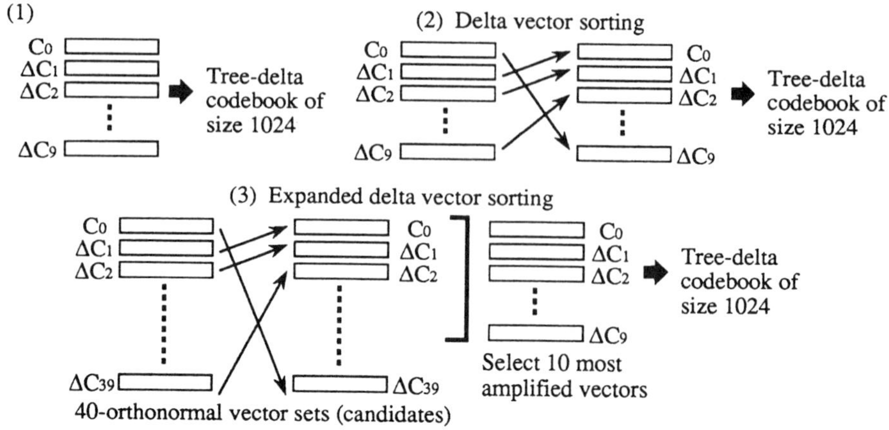

Figure 6. Delta vector sorting (Expansion)

CONCLUSION

A tree-delta codebook was presented as a structured stochastic codebook for an efficient CELP implementation. Also, a codebook adaptation method utilizing the structure of the tree-delta codebook was discussed. The performance of CELP with structured stochastic codebook is improved by the proposed delta vector sorting. Expanded delta vector sorting is applicable to other structured codebooks, such as the VSELP codebook.

REFERENCES

[1] B.S. Atal and M.R. Schroeder, "Stochastic Coding of Speech Signals at Very Low Bit Rates," *Proc. ICC*, pp. 1610-1613, May 1984.

[2] W.B. Kleijn et al, "Fast Methods for the CELP Speech Coding Algorithm," *IEEE Trans. on ASSP*, vol. 38, No.8, pp. 1330-1342, August 1990.

[3] G. Davidson and A. Gersho, "Complexity Reduction Methods for Vector Excitation Coding," *Proc. ICASSP*, pp. 3055-3058, April 1986.

[4] J-P. Adoul et al, "Fast CELP Coding Based on Algebraic Codes," *Proc. ICASSP*, pp. 1957-1960, April 1987.

[5] J.P. Campbell et al, "An Expandable Error-Protected 4800 BPS CELP Coder (U.S. Federal Standard 4800 BPS Voice Coder)," *Proc. ICASSP*, pp. 735-738, May 1989.

[6] I. Gerson and M. Jasiuk, "Vector Sum Excited Linear Prediction (VSELP) Speech Coding at 8 Kb/s," *Proc. ICASSP*, pp. 461-464, April 1990.

[7] M. Johnson and T. Taniguchi, "Pitch-Orthogonal Code-Excited LPC," *Proc. GLOBECOM*, pp. 542-546, Dec. 1990.

[8] T. Taniguchi et al, "Pitch Sharpening for Perceptually Improved CELP, and the Sparse-Delta Codebook for Reduced Computation," *Proc. ICASSP*, pp. 241-244, May 1991.

28

EFFICIENT MULTI-TAP PITCH PREDICTION FOR STOCHASTIC CODING

Dale Veeneman and Baruch Mazor

GTE Laboratories Incorporated
Waltham, MA 02254, USA

INTRODUCTION

In addition to the codebook excitation, the pitch or long-term predictor is of critical importance in determining the quality of the reconstructed speech in stochastic or CELP coding. Of equal concern is that when the pitch predictor is used in a "closed-loop" configuration, it has a high computational complexity and consumes, with the codebook search, a major portion of the coder's computational requirement. While a higher-order predictor (e.g., 3 adjacent taps) provides improved performance (primarily because of the implicit interpolated non-integer value for the effective lag), it also requires an increase in complexity (especially when performing a closed-loop analysis) and an increase in bit-rate. However, it is possible to reduce the complexity of a 3-tap filter to be nearly comparable to a 1-tap filter and use a moderate increase in bit-rate to gain an increase in performance.

COMPUTATIONAL COMPLEXITY

The general form of the long-term predictor is given as:

$$B(z) = 1 - \sum_{k=i}^{j} b_k z^{-(M+k)} \tag{1}$$

where the lag is M, the predictor coefficients are b_k and the number of taps are determined by i and j (i=j=0 is a one-tap predictor and i=–1, j=1 is a three-tap predictor). The closed-loop long-term prediction search operates much the same as that for the codebook excitation. For each of a series of lags, optimal coefficients are calculated that minimize the mean weighted squared error between the input speech and the synthetic speech resulting from the long-term filter using that lag. The lag and set of coefficients that give the smallest error for that frame is then used. The filtering process for the synthetic speech uses the same weighted LPC formant filter that is used in the subsequent codebook excitation search. Because the search proceeds through adjacent lags, the LPC filtering may be efficiently performed in a one-tap predictor by a recursion based on the past filtered result with an end-point calculation using the LPC impulse response. Then two correlations are needed (a cross-correlation and an energy calculation) to determine the optimum coefficient and weighted error for that lag. For a three-tap predictor, because the taps are adjacent, the filtering requires *no* additional calculations over the first-order method (results for past lags are used). Of the nine correlations needed, five are equal to previous results, therefore only four new correlations per lag are required (twice the number for the 1-tap case). Thus the 3-tap search requires a little more than twice

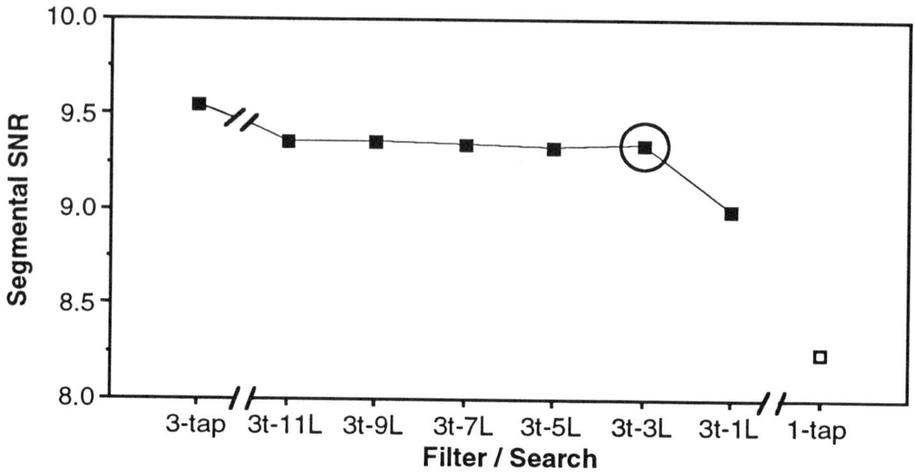

Figure 1. Performance of different 3-tap searches after a 1-tap search.

the operations as the 1-tap (roughly 250 *vs.* 122 multiply/add operations per lag). The coefficients may be calculated from the correlations in a closed form and then quantized or (as we do) the vector quantized coefficients may be determined by applying the correlations and the coefficient codebook entries to the error expression.

Despite the performance improvement of the 3-tap prediction, the computational cost is still quite high. Experiments comparing the 1-tap and 3-tap solutions revealed that often the optimum lags for each of the methods were close; the improvement was due to the addition of the adjacent taps. Therefore, a method was devised that performs a 1-tap search over the entire range of lags and then a 3-tap search over a neighborhood of lags surrounding the best 1-tap lag. Neighborhoods of 11 down to 1 lag were tested with the results shown in Figure 1. Listening tests revealed no perceived difference in quality between the full 3-tap search and the 1-tap search followed by the 3-lag neighborhood 3-tap search (further reductions gave noticeable distortion). However, the additional complexity required for the limited 3-tap search over the 1-tap search is quite small (less than a 7% increase). Thus for little additional complexity a significant improvement in performance can be realized.

WHEN THE LAG IS LESS THAN THE FRAME SIZE

Another complexity issue concerns the persistent problem with the implementation of the closed-loop long-term analysis when the filter lag is less than the size of the predictor frame. For lags greater than the frame size, past known data is used to predict the current frame. However, at lags less than the frame size, the "exact" solution for the one-tap case requires the coefficient to be squared for the repeated part (cubed if it repeats again), because the repeated part will re-apply the coefficient to data already scaled by that coefficient. This requires a non-linear solution for the coefficient and for the 3-tap case, the problem is compounded greatly. The popular alternative is to redefine the operation of the long-term filter. For lags greater than the frame length, the filter operates as usual, but for lags less

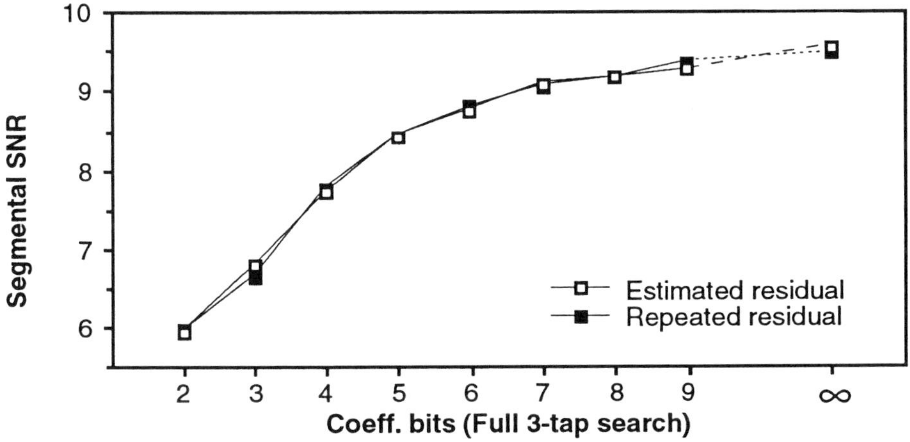

Figure 2. Performance of estimated and repeated residual methods.

than the frame length, the past synthetic residual is *repeated* as often as necessary and the coefficients are applied equally to the entire frame (this has been called a virtual search or an adaptive codebook method [1]). The disadvantage of this method is that computational efficiency is reduced when the adjacent sample recursion is lost for lags less than the frame length.

We have derived another alternative that retains the usual definition of the long-term filter and uses the inverse (LPC) filtered residual of the input speech for the repeated part of the past synthetic residual during the analysis (only for lags less than the frame size). This allows the adjacent sample recursion to be used for the entire range of lags. The motivation is that the unknown synthetic residual and the input speech residual should not be too dissimilar. Any minor inconsistencies will then be corrected in the following code excitation stage. Note that the input residual is only used in the analysis to calculate the coefficients. The transmitter and receiver use the same synthetic residual as filter memories (i.e., they both synthesize the same speech). Even though the analysis no longer equals the synthesis, listening tests and segmental SNR (see Figure 2) revealed no difference between the two methods. Moreover, the described *estimated* residual method is simpler to implement and computationally more efficient.

COMPARISON TO INTERPOLATED LAGS

We have found that the quantized multi-tap predictor compares favorably with the single-tap predictors that use interpolated, non-integer lags. The interpolated lags were calculated using the method of [2], which is to increase the sampling frequency (by inserting zeros), low-pass filtering with a Hamming-weighted $\sin(x)/x$ interpolating filter, and then down-sampling about the lag of choice. We used uniform interpolating factors of 1 (none), 2, 4, and 8.

A comparison of 1-tap, 3-tap and 8th-order interpolated lag prediction filters is shown in Figure 3 (the method of the *repeated* residual for lags less than the frame size was used in all cases). The comparison is the segmental SNR performance *vs.*

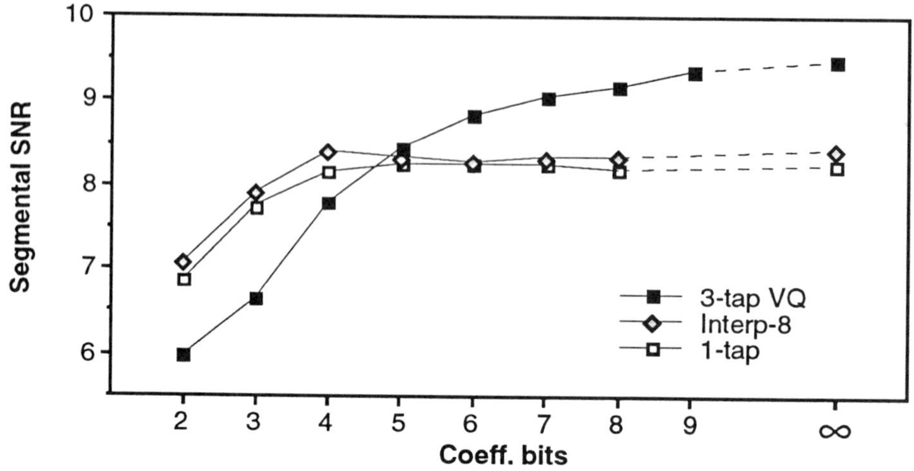

Figure 3. Performance of 3-tap, 8th order interpolated lag and 1-tap predictors.

the number of bits used to represent the coefficients. The coder is fully quantized about a 6 kb/s base rate (the number of pitch predictor coefficient bits is changed independently) and the data was a mix of male and female speakers under a wide variety of *telephone* conditions. Note that 2 or 3 additional bits for the 3-tap filter beyond the traditional 4 or 5 bit coefficient gives an advantage over that of the interpolated filter (which also requires from 1 to 3 additional bits for the additional lags).

The performance was also compared using the prediction gain as a measure (see Figure 4). The pitch prediction was performed on the LPC residual (quantized 10th order autocorrelation every 25 msec), with the pitch parameters calculated every 6.25 msec. (50 sample frame length). Only frames with prediction gain greater than 1.2

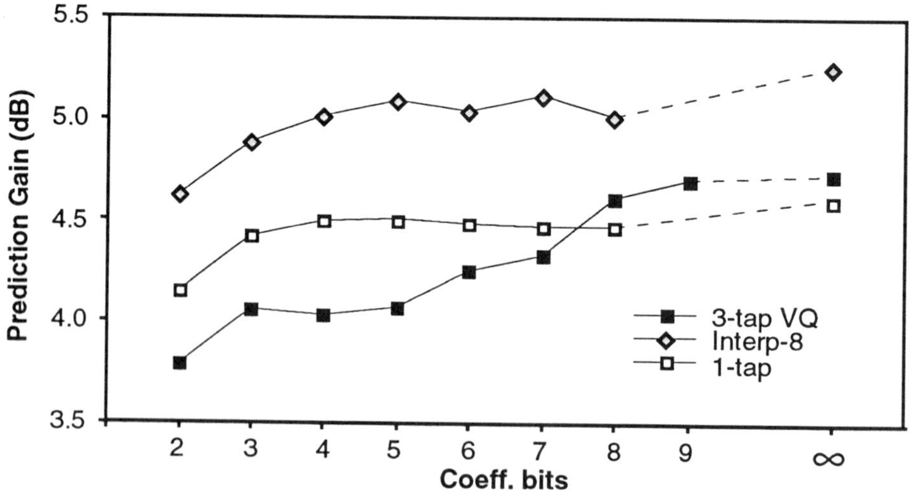

Figure 4. Prediction gain performance of pitch predictors (for frames > 1.2 dB).

Table 1. Segmental SNR comparison of pitch prediction filters for frames with prediction gain greater than 1.2 dB.

Filter	SegSNR	% of frames
3-tap	14.03 dB	36%
Interp (8)	12.40 dB	33%
1-tap	11.16 dB	30%

dB were used in the average (as in [2]). It is interesting that prediction gain favors the interpolated filter (as reported in [2]) and segmental SNR favors the multi-tap filter (as reported in [3], where a *two*-tap filter was used). When all frames were used in the average prediction gain, the interpolated predictor and the 3-tap predictor gave approximately equal performance.

Tests were conducted to determine if the reason for the prediction gain results was that the interpolated lag filter out-performs the 3-tap filter during voiced frames. Table 1 gives the segmental SNR results for frames that had a *prediction gain* greater than 1.2 dB. Note than the 3-tap filter gave more frames with prediction gain greater than 1.2 dB and that the average segmental SNR was significantly higher for those frames. In addition, in informal listening tests the 3-tap filter was preferred.

An important consideration is the computational cost. With an interpolated lag filter the interpolation is not only required during the pitch search, but also during the pitch filter memory subtraction (before the code excitation) and also in the decoder. As already mentioned, the 3-tap filter requires very little additional complexity. The cost in increased bit rate is about equal. The 3-tap filter may use 2 extra bits (from 5 to 7), while the interpolated lag filter may use from 1 (non-uniform) to 3 extra bits.

CONCLUSION

We have described a multi-tap pitch predictor method with vector quantized coefficients that exhibits performance gains superior to traditional single-tap predictors at moderate bit rates. Significantly, this performance increase requires only a minor increase in the computational complexity. Moreover, the minor increase in bit rate for the 3-tap predictor is more than compensated in quality and allows the bits to be recovered elsewhere.

REFERENCES

[1] J. P. Campbell, Jr., V. C. Welch, and T. E. Tremain, "The new 4800 bps voice coding standard," *Proc. Military & Govt. Speech Tech '89*, pp. 735-737, Nov. 1989.

[2] P. Kroon and B. S. Atal, "Pitch Predictors with High Temporal Resolution," *Proc. IEEE Int. Conf. Acoust., Speech, Signal Processing*, pp. 661-664, Apr. 1990.

[3] J. S. Marques, *et. al.*, "Improved Pitch Prediction with Fractional Delays in CELP Coding," *Proc. IEEE Int. Conf. Acoust., Speech, Signal Processing*, pp. 665-668, Apr. 1990.

29

QR FACTORIZATION IN THE CELP CODER

Przemyslaw Dymarski † and Nicolas Moreau ‡

† Technical University of Warsaw
ul. Nowowiejska 15/19, 00-665 Warsaw, POLAND

‡ Telecom Paris, Dept Signal
46 rue Barrault, 75634 Paris Cedex 13, FRANCE

INTRODUCTION

In most recently proposed speech coders at bit rates between 4.8 and 16 kbit/s, the synthetic speech signal is obtained by filtering a synthetic excitation signal through an all-pole filter, that models the vocal tract spectral characteristics. Defining the excitation signal has been, and still is, an active field of research. Multipulse coders, CELP coders, regular-pulse coders , etc ... have been quite successful and they basically differ by the structure of their excitation. These coders usually include a long-term predictor, which can be seen as an adaptive codebook containing past excitation. In a general way, the excitation vector $\hat{e}$ may be modeled as a linear combination of K signals, originated from K codebooks and multiplied by K associated gains

$$\hat{e} = \sum_{k=1}^{K} g_k c_k^{j(k)} \tag{1}$$

where $c_k^{j(k)}$ denotes the j(k)-th column vector of the codebook C_k as shown in Figure 1. This general case corresponds, for example, to the multistage CELP coder presented in [1]. This model can be used for all of the above-mentioned coders, if we define the codebooks and apply appropriate constraints on indices and/or gains. The indices $j(k)$ and the gains g_k are computed in order to minimize the Euclidean distance between the original perceptual signal p and the synthetic perceptual signal $\hat{p}$. For given indices $j(1) \cdots j(K)$, computing the gains is a classical linear least squares estimation problem [2]. This minimization problem exhibits two particular properties that make it difficult to apply the classical information and signal theory results. First, there are perceptual signals involved, related to the codebook signals by a filtering operation. Second, a small number of vectors (e.g. 2 or 3) is selected from codebooks consisting of many vectors (e.g. 256 or 512).

This paper deals with the problem of this minimization. We investigate several algorithms that construct the synthesis filter's input in the CELP coder.

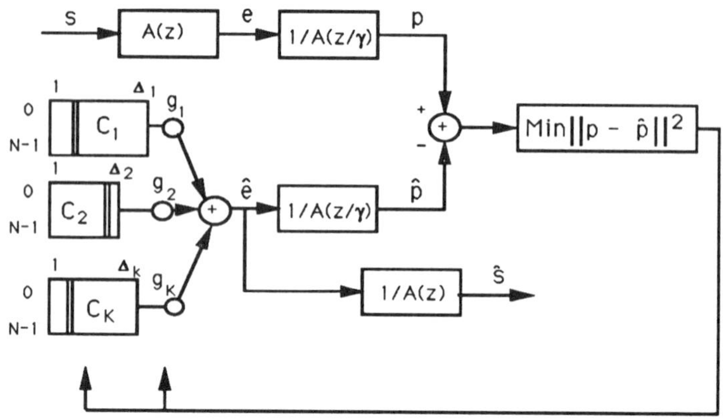

Figure 1: **Multistage CELP coder**

Then all these algorithms are evaluated with respect to their computational cost and SNR improvement using realistic values for the parameters. The bit rate chosen for this test is around 9 kbit/s. The problem of the excitation codebook determination is not discussed in this paper.

LOCALLY OPTIMAL ALGORITHMS

The Euclidean distance between the original and the synthetic speech signal, expressed at the perceptual level $E = ||p - \hat{p}||^2$, must be minimized. The synthetic perceptual vector $\hat{p}$ depends not only on the unknown parameters $j(1) \cdots j(K)$ and $g_1 \cdots g_K$ but also on the excitation signal of the preceding frames. We obtain

$$E = ||(p - \hat{p}^0) - \sum_{k=1}^{K} g_k H c_k^{j(k)}||^2 \tag{2}$$

where H is the lower triangular Toeplitz matrix obtained from the impulse response of the perceptual filter $1/A(z/\gamma)$ in the current frame. Let us simplify the notation using p for the perceptual vector minus the contribution $\hat{p}^0$ from the preceding frames and $\hat{p}$ the last term in (2). From the excitation codebooks C_k, we obtain the filtered codebooks F_k with $F_k = HC_k$. They are composed of column vectors f_k^j. For computing the optimal excitation, we have to select the subspace $f_1^{j(1)} \cdots f_K^{j(K)}$ maximizing the norm $||\hat{p}||^2$ where $\hat{p}$ is the orthogonal projection of p on this subspace. The gains $g_1 \cdots g_K$ are obtained by solving the normal equations. With K distinct codebooks of dimension $N * \Delta_k$, $\prod_k \Delta_k$ subspaces are possible. Therefore computationally-wise, the optimal algorithm is far too complex. With the well-known standard algorithm, the indices and gains are computed in a recursive fashion. At each step, the energy $\alpha^j = < f_k^j, f_k^j >$ of the filtered vectors and the crosscorrelations $\beta^j = < f_k^j, p - \sum_{n=1}^{k-1} g_n f_n^{j(n)} >$ are evaluated ($< x, y >$ denotes the inner product of two vectors). We choose the index $j(k)$ minimizing the angle between the modeling error and f_k^j or maximizing $(\beta^j)^2/\alpha^j$. The corresponding gain is $\beta^{j(k)}/\alpha^{j(k)}$. Two classical methods are available to reduce the suboptimality of this algorithm. The first is to globally optimize the gains at the end of the minimization procedure. The second is to optimize the gains at each step.

We have proposed [3] a third way to reduce the suboptimality of the standard algorithm. At the kth step the already determined subspace $f_1^{j(1)} \cdots f_{k-1}^{j(k-1)}$ of dimension $k-1$ is augmented with the vector $f_k^{j(k)}$ maximizing the norm of the projection of p on the k dimensional subspace spanned by $f_1^{j(1)} \cdots f_{k-1}^{j(k-1)} f_k^j$. This is still too complex but this computational cost is reduced if an orthogonal basis is progressively created in this subspace. It is sufficient to orthogonalize the codebook F_k relative to the $k-1$ vectors chosen previously or to orthogonalize the codebooks $F_k \cdots F_K$ relative to the vector chosen at the $(k-1)^{th}$ step.

The orthogonalization of one vector f^j relative to a normalized vector q is described by

$$f_{orth}^j = f^j - < f^j, q > q = f^j - r^j q \tag{3}$$

where the crosscorrelation r^j is the component of f^j on q. This orthogonalization is also equivalent to the projection of f^j on the subspace orthogonal to q. This projection can be expressed by the square matrix $P = I - qq^t$ since

$$f_{orth}^j = f^j - qq^t f^j = P f^j \tag{4}$$

Let q_k denote the normalized vector selected in the codebook F_k, orthogonalized $(k-1)$ times, r_k^j the component of vectors f^j on q_k and P_k the corresponding projection matrix. At the kth step, the vectors $f_{orth(k)}^j$, orthogonalized $(k-1)$ times, are given by

$$f_{orth(k)}^j = f_{orth(k-1)}^j - r_{k-1}^j q_{k-1} = f^j - \sum_{n=1}^{k-1} r_n^j q_n \tag{5}$$

We have also

$$f_{orth(k)}^j = P_{k-1} P_{k-2} \cdots P_1 f^j \tag{6}$$

Maximizing the norm of the projection in the subspace spanned by the orthogonal basis $f_1^{j(1)} f_{orth(2)}^{j(2)} \cdots f_{orth(k-1)}^{j(k-1)} f_{orth(k)}^j$ consists of choosing the vector maximizing $(\beta_k^j)^2/\alpha_k^j$ with

$$\alpha_k^j = < f_{orth(k)}^j, f_{orth(k)}^j > \tag{7}$$

$$\beta_k^j = < f_{orth(k)}^j, p - \sum_{n=1}^{k-1} g_n f_n^{j(n)} > = < f_{orth(k)}^j, p > \tag{8}$$

The explicit computation of the orthogonalized vectors is not necessary. It is sufficient to update the energies α_k^j and the crosscorrelations β_k^j. We obtain

$$\alpha_k^j = \|f_{orth(k-1)}^j - r_{k-1}^j q_{k-1}\|^2 = \alpha_{k-1}^j - (r_{k-1}^j)^2 \tag{9}$$

$$\beta_k^j =< f_{orth(k-1)}^j - r_{k-1}^j q_{k-1}, p >= \beta_{k-1}^j - r_{k-1}^j \frac{\beta_{k-1}^{j(k-1)}}{\sqrt{\alpha_{k-1}^{j(k-1)}}} \tag{10}$$

For updating energies and crosscorrelations, it is sufficient to know the crosscorrelations r_{k-1}^j at each step. We have shown [4] that these crosscorrelations can be obtained recursively

$$r_{k-1}^j = \frac{1}{\sqrt{\alpha_{k-1}^{j(k-1)}}}[< f^j, f_{k-1}^{j(k-1)} > - \sum_{n=1}^{k-2} r_n^j r_n^{j(k-1)}] \tag{11}$$

This evaluation is done for all vectors belonging to the filtered codebooks $F_k \cdots F_K$. Since the crosscorrelations r_n^j are the components of the vectors f^j on the new basis $q_1 \cdots q_k$, the preceding computation corresponds to the beginning of the QR factorization of the matrix composed of the vectors f^j without the explicit computation of Q. The classical QR factorization is performed only for the vectors $f_1^{j(1)} \cdots f_K^{j(K)}$ since

$$\left[\begin{array}{cccc} f_1^{j(1)} & . & . & f_K^{j(K)} \end{array} \right] = \left[\begin{array}{cccc} q_1 & . & . & q_K \end{array} \right] \left[\begin{array}{ccccc} r_1^{j(1)} & r_1^{j(2)} & . & . & r_1^{j(K)} \\ 0 & r_2^{j(2)} & . & . & r_2^{j(K)} \\ . & . & . & . & . \\ . & . & . & . & r_{K-1}^{j(K)} \\ 0 & . & . & 0 & r_K^{j(K)} \end{array} \right] \tag{12}$$

but it is extended (i.e. the crosscorrelations r_n^j are calculated) to the other columns of the codebooks $F_1 \cdots F_K$. The equations (9) - (11) lead to the Recursive Modified Gram-Schmidt (RMGS) algorithm presented in [4]. For low bit rate coders, only two excitation codebooks generally are used (an adaptive and a stochastic codebook) with one vector chosen in each codebook. In this case, $K = 2$ and the second term in (11) disappears. The RMGS algorithm reduces to an elementary form as described in [5].

Let us note that the explicit computation of the filtered codebooks is not necessary since the preceding formulae can be expressed using only the vectors c^j, because of the well-known transformations $< f^j, f^j >= (c^j)^t H^t H c^j$ and $< f^j, p >= (c^j)^t (H^t p)$. This formulation is widely used as the computational cost is reduced when special structures are imposed on the excitation codebooks (e.g. sparse, algebraic codebooks ...).

Closed-loop quantization of the gains is then introduced. These gains can be computed relatively to both the original codebook vectors $f_k^{j(k)}$ and the orthogonalized codebook vectors $q_k = f^{j(k)}_{orth(k)}/||f^{j(k)}_{orth(k)}||$. We propose a new coding method based on the special distribution of the gains $\bar{g}_k$ relative to the orthogonal vectors. For this kind of gains the following property is satisfied

$$||\hat{p}||^2 = \sum_{k=1}^{K} \bar{g}_k^2 = \bar{g}_1^2(1 + (\frac{\bar{g}_2}{\bar{g}_1})^2(1 + \cdots (\frac{\bar{g}_K}{\bar{g}_{K-1}})^2)) \tag{13}$$

This suggests the indirect (adaptive) coding of the gains, relative to the value $||\hat{p}||^2$. Instead of the modeled perceptual vector $\hat{p}$ we use the original perceptual vector p. The norm $||p||^2$ may be calculated and coded less frequently (for example once per 20 ms) than the gains (for example once per 5 ms). Thus the first gain is expressed in the following way $\bar{g}_1^2 = \lambda||p||^2$ and the coefficient λ is quantized. Then the ratios $\bar{g}_2/\bar{g}_1 \cdots \bar{g}_K/\bar{g}_{K-1}$ are coded using nonuniform quantizers.

At the synthesis part, since only non-orthogonalized excitation codebooks are available, we have to perform a new QR factorization of the matrix $f_1^{j(1)} \cdots f_K^{j(K)}$ with no extension to the other vectors. The computational cost is not of the same order of magnitude at the analysis and synthesis levels. The typical ratio is about 100.

SIMULATION RESULTS

These algorithms are evaluated with respect to their computational cost and SNR improvement. The experiments were run in the following way. The short term predictor is updated every 20 ms (160 samples for 8 kHz sampling frequency) by a 8th order LPC analysis based on Schur's algorithm. The log area ratios are coded with 36 bits, which corresponds to a bit rate of 1.8 kbit/s. The excitation signal is modeled using K vectors every 5 ms ($N = 40$). The first vector is extracted from an adaptive codebook consisting of $\Delta_1 = 128$ vectors and the $K - 1$ remaining vectors are selected from a stochastic codebook with $\Delta_2 = 128$ vectors, populated with gaussian random variables.

Every 20 ms, the energy of speech signal at the perceptual level is coded with 5 bits. Every 5 ms, the coefficient λ and the gain ratios are coded with 4 bits, the indices with 7 bits. All coding tables are computed using the LBG algorithm. The sign of the first gain must be transmitted. The bit rate for the excitation signal is therefore $0.45 + 2.2 * K$ kbit/s which yields 8.85 kbit/s for $K = 3$.

To give an order of magnitude for the computational cost, we evaluate the number of multiplications/ accumulations in Mflops (10^6 floating point operations per second). Using the properties of the Toeplitz adaptive codebook (a sample shift between 2 adjacent vectors), the iterative standard algorithm needs 6.8 Mflops for $K = 3$. Some details about this evaluation are given in Table 1.

LPC analysis, Perceptual	0.2 Mflops
Adaptive codebook filtering	0.3 Mflops
Energy α^j	0.29 Mflops
Crosscorrelation β^j	1.02 Mflops
Stochastic codebook filtering	2.05 Mflops
Energy α^j	0.25 Mflops
Crosscorrelation β^j	1.02 Mflops
Update of r_k^j	1.02 $(K-2)$ Mflops
$j(k)$ and rest	0.17 K Mflops

Table 1: **Computational cost for the iterative standard algorithm**

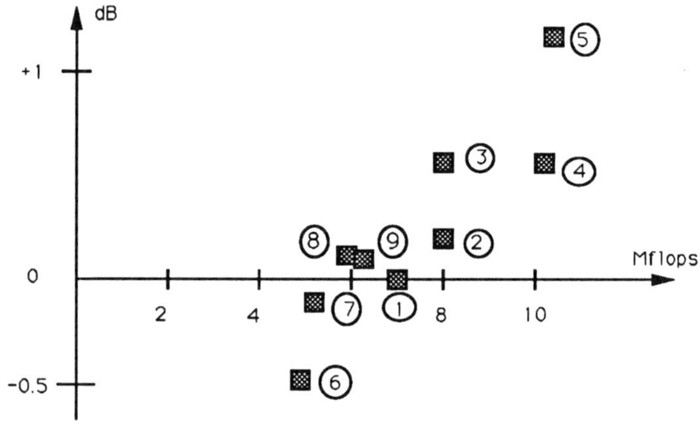

Figure 2: **Computational cost and SNR improvement**

The algorithms are tested on 4 sentences uttered by two female and two male speakers, about 24 seconds of total speech. Figure 2 shows the results.

Case 1 corresponds to the iterative standard algorithm, case 2 to the algorithm with gain optimization at each step and case 3 to the RMGS algorithm. In case 4 with the standard algorithm and case 5 with the RMGS algorithm, the adaptive and stochastic codebooks are grouped together and the coder can choose $K = 3$ vectors from this mixed codebook. The bit rate is thus increased by 600 bits/s and the results cannot be compared with the other cases. A more detailed examination of this mixed codebook approach shows that there is a slight SNR improvement even at the same bit rate [3]. The computational cost may be reduced in several ways. We give results only for two classical cases. The first one consists in forcing the stochastic codebook to be Toeplitz [6] (case 6 with the standard algorithm and case 8 with the RMGS algorithm). In the second one, we suppress the filtered codebooks and force the matrix $H^t H$ to be Toeplitz, a widely used modification [7] (case 7 with the standard algorithm and case 9 with the RMGS algorithm).

CONCLUSION

For defining the excitation signal in a multistage CELP coder, we propose a locally optimal algorithm based on QR factorization.

Simulations of a 9 kbit/s 3-stage coder show that this algorithm offers higher SNR (0.5 dB) than the standard iterative algorithm with small additional computational cost (0.5 Mflops) but informal listening tests indicate no significant improvement of speech quality in this case.

The advantages of the proposed algorithm are more evident with greater or variable number of stages, e.g. for an embedded CELP coder for wideband speech coding as described in [8].

REFERENCES

1. G. Davidson and A. Gersho "Multiple Stage Vector Excitation Coding of Speech Waveforms" Proc. Int. Conf. Acoust., Speech, Signal Processing, pp. 163-166, 1988

2. G. Golub and C. Van Loan "Matrix Computations" Johns Hopkins University Press, 1983 (Second Edition 1989)

3. N. Moreau and P. Dymarski "Mixed Excitation CELP Coder" Proc. Eurospeech, pp. 322-325, 1989

4. P. Dymarski, N. Moreau and A. Vigier "Optimal and Sub-optimal Algorithms for Selecting the Excitation in Linear Predictive Coders" Proc. Int. Conf. Acoust., Speech, Signal Processing, pp. 485-488, 1990

5. J.H. Yao, J. Shynk and A. Gersho "Low-Delay Vector Excitation Coding of Speech at 8 Kbit/s" Proc. Globecom'91

6. D. Lin "Speech Coding Using Efficient Pseudo-Stochastic Block Codes" Proc. Int. Conf. Acoust., Speech, Signal Processing, 1987

7. I. Trancoso and B. Atal "Efficient Procedures for Finding the Optimal Innovation in Stochastic Coders" Proc. Int. Conf. Acoust., Speech, Signal Processing, pp. 2375-2378, 1986

8. A. Le Guyader, B. Lozach and N. Moreau "Embedded Algebraic CELP Coders for Wideband Speech Coding" Proc. EUSIPCO-92, Vol. 1 pp. 527-530, 1992

30

EFFICIENT FREQUENCY-DOMAIN REPRESENTATION OF LPC EXCITATION

Sunil K. Gupta and Bishnu S. Atal

AT&T Bell Laboratories
Murray Hill, New Jersey 07974, USA

INTRODUCTION

Efficient representation of LPC excitation signal is of utmost importance in predictive coding systems for achieving high quality speech at low bit rates. In this paper, we present a method for obtaining an efficient parametric representation of the LPC excitation signal for voiced speech in the frequency domain that takes advantage of the nonuniform spacing of critical bands [1] in the auditory system. In current analysis/synthesis systems [2,3], a significant portion of the available bits is used to represent the excitation signal in order to reproduce its detailed structure which is very complicated. The method presented in this paper aims to preserve only those details in the LPC excitation signal which are necessary to produce synthetic speech without audible distortion.

A segment of the LPC excitation signal with a duration of N samples, represented as a Fourier series, requires N/2 sinusoidal components uniformly spaced along the frequency axis for its exact reproduction. In the sparse frequency-domain representation described in this paper, the LPC excitation signal is represented in terms of only a few non-orthogonal time-windowed sinusoidal basis functions.

The technique presented in this paper leads to a few parameters describing each pitch-cycle of the excitation waveform that vary smoothly from one pitch-cycle to the next during slowly evolving segments of voiced speech. For such segments, it is further possible to update the parameters every 20-30ms. These steps are shown in Fig. 1. In this scheme, one pitch-cycle of LPC excitation is extracted every 20-30ms and is analyzed using the sparse representation. The parameters for the intermediate pitch-cycles are then generated by interpolation [4].

The sparse frequency-domain representation could lead to reduction in the bit-rate required for transmitting LPC excitation parameters. This reduction, however, depends strongly on the coding strategy and the quantization characteristics of the parameters. Finding appropriate quantization schemes is beyond the scope of this paper.

FREQUENCY-DOMAIN REPRESENTATION

Let $u(n),\ 0 \leq n \leq N-1$, denote a period of LPC excitation. The signal $u(n)$ can be

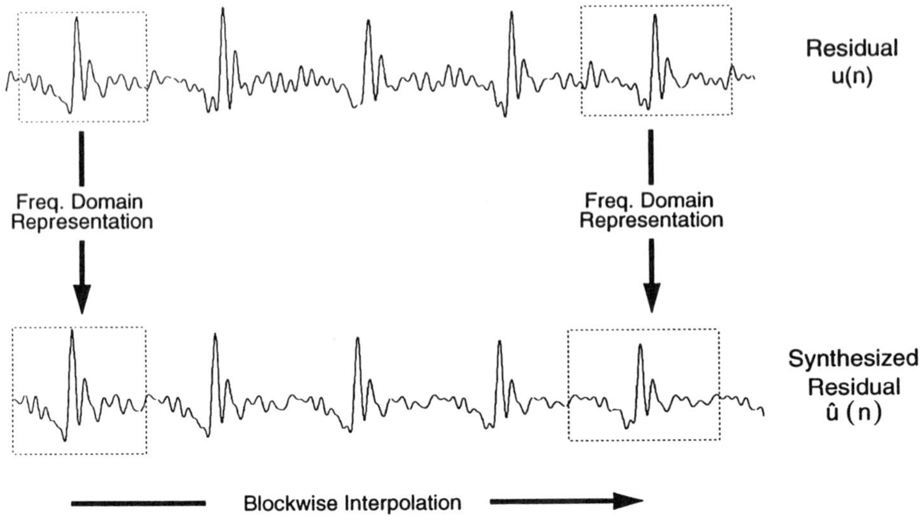

Fig. 1. Sparse Frequency-Domain Representation and Blockwise Interpolation.

represented exactly by means of the Fourier series:

$$u(n) = \sum_{k=0}^{N-1} a_k \cos(k\omega_0 n) + \sum_{k=1}^{N-1} b_k \sin(k\omega n), \quad 0 \leq n \leq N-1, \tag{1}$$

where ω_0 is the fundamental frequency, and a_k, b_k are the Fourier coefficients. Due to the symmetry properties of the Fourier series representation, the number of distinct parameters in the above equation is only N.

In the sparse representation, we approximate a period of the excitation signal in terms of a small set of time-windowed basis functions. That is,

$$\hat{u}(n) = \sum_{k=0}^{K} a'_k w_k(n) \cos(\omega_k n) + \sum_{k=1}^{K} b'_k w_k(n) \sin(\omega_k n), \quad 0 \leq n \leq N-1, \tag{2}$$

where K is the number of basis functions selected for the sparse representation ($K \leq N$); $w_k(n), k = 0, \ldots, K$, are the window functions; and a'_k, b'_k are the coefficients for the sparse representation. The frequencies $\omega_k, k = 0, \ldots, K$ are uniformly spaced at low frequencies and logarithmically spaced at high frequencies. In (2), let

$$\Psi_k(n) = w_k(n) \cos(\omega_k n), \tag{3}$$

and

$$\Phi_k(n) = w_k(n) \sin(\omega_k n). \tag{4}$$

The mean-squared error E can be written as

$$E = \sum_{n=0}^{N-1} \left[u(n) - (\sum_{k=0}^{K} a'_k \Psi_k(n) + \sum_{k=0}^{K} b'_k \Phi_k(n)) \right]^2, \quad b_0^{prime} = 0. \tag{5}$$

Computing the partial derivatives with respect to the parameters a'_i and b'_i and equating them to zero, we obtain

$$\sum_{k=0}^{K}\left[a'_k\sum_{n=0}^{N-1}\Psi_i(n)\Psi_k(n)+b'_k\sum_{n=0}^{N-1}\Psi_i(n)\Phi_k(n)\right]=\sum_{n=0}^{N-1}u(n)\Psi_i(n),\quad 0\le i\le K,$$

$$\sum_{k=0}^{K}\left[b'_k\sum_{n=0}^{N-1}\Phi_i(n)\Phi_k(n)+a'_k\sum_{n=0}^{N-1}\Phi_i(n)\Psi_k(n)\right]=\sum_{n=0}^{N-1}u(n)\Phi_i(n),\quad 0\le i\le K. \tag{6}$$

The above simultaneous linear equations are solved to obtain the parameters a'_k and b'_k. The magnitude c'_k and phase ϕ'_k are given by

$$\begin{aligned} c'_k &= [(a'_k)^2+(b'_k)^2]^{1/2}, \quad 0\le k\le K,\\ \phi'_k &= \arctan[b'_k/a'_k], \quad 0\le k\le K. \end{aligned} \tag{7}$$

Frequency-Domain Sampling

Since the frequency selectivity of the human ear is nonuniform and decreases at high frequencies, it is possible to use a relatively sparse spectral representation of the LPC excitation at high frequencies without introducing audible distortion in the reconstructed speech signal. At low frequencies, however, the excitation signal must be represented very accurately. To achieve this, the low-frequency sinusoidal components are uniformly spaced and high-frequency components are logarithmically spaced. That is,

$$\omega_k=\begin{cases} k\omega_0, & 0\le\omega_k\le\omega_M,\\ \alpha\omega_{k-1}, & \omega_M<\omega_k\le\omega_c, \end{cases} \tag{8}$$

where ω_M is the cut-off frequency below which the sinusoidal components are equally spaced and ω_c is the bandwidth of the input signal. The parameter α determines the spacing between adjacent components for frequencies above ω_M. Increasing the spacing parameter α results in an increasingly sparse spectral representation at frequencies above ω_M. For approximately one-third octave spacing, the parameter α is 1.25. The number of frequency samples K is determined such that $\omega_K\le\omega_c$. It is clear from (8) that for high pitched voices (e.g. children and females), the number of components K will be much smaller than for the relatively low pitched voices (e.g. males).

Note that the above method provides different number of components as the pitch is varied. It is possible, however, to obtain a fixed number of components for each pitch period by specifying the number of components K and varying the spacing parameter α.

Selection of Window Functions

Due to the increase in spacing between the adjacent components for frequencies above ω_M, a period of LPC excitation is represented by fewer parameters in the sparse representation than in the exact Fourier representation. It is important to ensure that the new representation still spans the complete bandwidth of the input speech signal. Any frequency band that is not present in the reconstructed excitation signal produces synthetic speech that has tonal quality. In our method, we vary the time-width of the basis functions by multiplying with a window function $w_k(n)$ in (2), since this is equivalent to varying the bandwidth of the corresponding basis function. The time-width of the window functions is made inversely proportional to the frequency range $\triangle\omega_k$ that must be spanned by each basis function. $\triangle\omega_k$ is given by

$$\triangle\omega_k = \begin{cases} \omega_{k+1} - \omega_k, & 1 \leq k \leq K-1, \\ max(\omega_c - \omega_K, \omega_K - \omega_{K-1}), & k = K. \end{cases} \tag{9}$$

For a rectangular window, the time-width N_k is defined as:

$$N_k = \lceil \frac{2\omega_c}{\triangle\omega_k} \rceil, \tag{10}$$

where $\lceil x \rceil$ represents the integer greater than or equal to x. For a Hanning window, the time-width is twice the value given by (10). Each time-window is placed symmetric relative to the center of the current pitch period. A further normalization step is performed so that the time-windows, for all the basis functions, have the same energy. Note that the variation in time-width of the window functions with frequency is similar to the approach used in a wavelet representation [8] and exploits the frequency selectivity of the human ear. Unlike in wavelet representation, however, we undersample the signal in the time-domain. For voiced speech, we have found that this undersampling does not introduce any audible distortion in the synthetic speech signal.

An example of the windowed basis functions is shown in Fig. 2. Figure 2(a) shows a time-windowed basis function using a rectangular window and its associated Fourier magnitude spectrum. We use rectangular windows below w_M to obtain accurate spectral information for low frequencies from the complete pitch cycle. This is essential to preserve the broad spectral characteristics of the glottal waveform. Examples of high frequency sinusoidal basis functions with a Hanning window are shown in Figs. 2(b)-(c). Note that for high frequencies, the basis functions span a much larger frequency region than at the low frequencies. As a consequence of the time-windowing, one must ensure that the main feature in the pitch-cycle waveform occurs in the center of the window. This is necessary to correctly reproduce the periodic behavior of voiced speech.

THE ANALYSIS/SYNTHESIS SCHEME

The sparse frequency-domain representation of the LPC excitation was implemented within an analysis/synthesis scheme. The sparse frequency-domain analysis is applied to each pitch-cycle of an upsampled version of the LPC residual signal to obtain the magnitudes c'_k and phases ϕ'_k as given by (7). Each pitch-cycle is defined around a pitch marker [5] to ensure that the main feature within the pitch-cycle occurs in the middle of the analysis window.

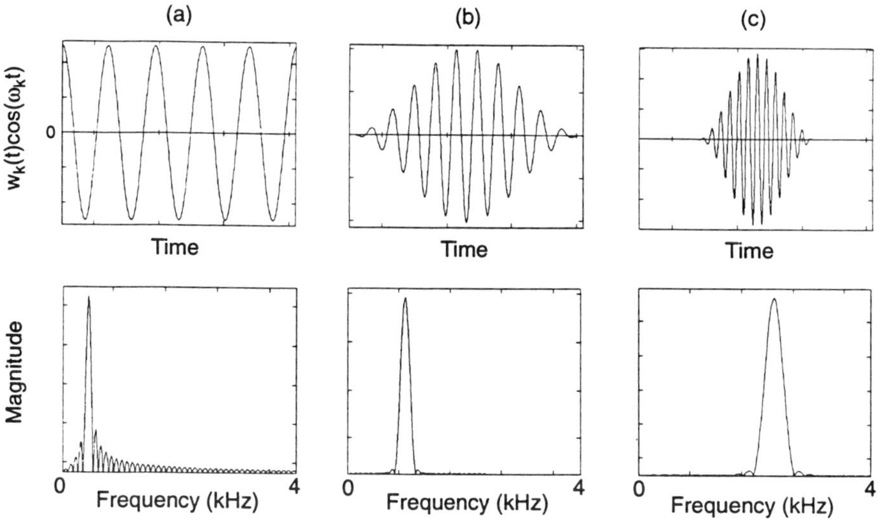

Fig. 2. Examples of Time-Windowed Sinusoidal Basis Functions and Their Associated Fourier Magnitude Spectra: (a) Rectangular Window and (b)-(c) Hanning Window.

In the synthesis phase, the LPC excitation signal is reconstructed from the magnitude and phase terms obtained during the analysis phase. The reconstructed LPC excitation signal is then passed through the synthesis filter to obtain reconstructed speech.

In Figs. 3(a)-(c), we show that for a pitch-cycle the original and the reconstructed LPC excitation signals as well as their associated Fourier magnitude and phase representations. Note that within a small time-window around the main feature (the peak) in the pitch-cycle, the waveform is reproduced well. The size of this window is inversely proportional to the largest spacing between the adjacent frequency components. It is also evident from Figs. 3(b)-(c) that the sparse representation results in smoothing of the magnitude and the phase information at high frequencies. Furthermore, our listening tests show that this smoothing is performed in a manner that does not introduce significant audible distortion in the reconstructed speech.

Figure 4 shows the plots of the energy, expressed in decibels, of the input speech signal and the energy of the difference signal between input and the reconstructed speech. Unvoiced and transitional segments have been replaced by the input speech signal. Careful examination of Fig. 4 shows that the signal-to-noise ratio (defined as the energy of the difference between the input signal and the error signal) is relatively low even though the subjective quality of the reconstructed speech was found to be very high.

Figure 5 compares the correlation ρ between successive pitch cycles for the input and the reconstructed LPC excitation signals. Note that for most part, correlation

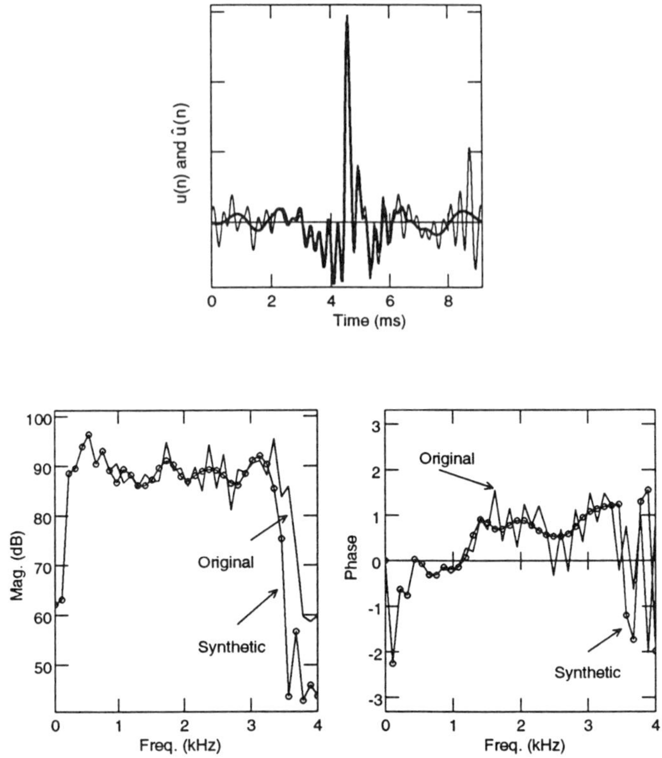

Fig. 3. An Example of the Original and Reconstructed Excitation Signals: (a) Time-Domain Representation (reconstructed signal is shown in bold), (b) Fourier Magnitude Representation, and (c) Fourier Phase Representation.

function for the synthetic signal lies above the correlation function for the input signal. This also suggests that the sparse frequency-domain representation removes that noise component of the signal which contributes least to the perceptual quality of speech.

A listening test was conducted to determine the mean-opinion-score (MOS) for the reconstructed speech when the sparse frequency-domain analysis is used only for the voiced segments. For this experiment, the cut-off frequency ω_M of 1000 Hz was used and the parameter α was set to 1.25. The MOS score was found to be 4.18. In comparison, the MOS score for the input speech was found to be 4.36. Furthermore, reference speech signals were generated by adding Gaussian noise to the speech signal, the amplitude of the noise being controlled by that of the speech signal. This system of obtaining reference speech signals is known as the Modulated Noise Reference Unit (MNRU) [6]. The MOS score for reference speech signals with 25 dB MNRU generated using the above method was also found to be 4.18. Note that only five or six frequency components are actually used to represent the 1-4 kHz frequency region.

It is important to note that the formant structure is removed from the speech signal before applying the frequency-domain representation. The analysis is also applicable

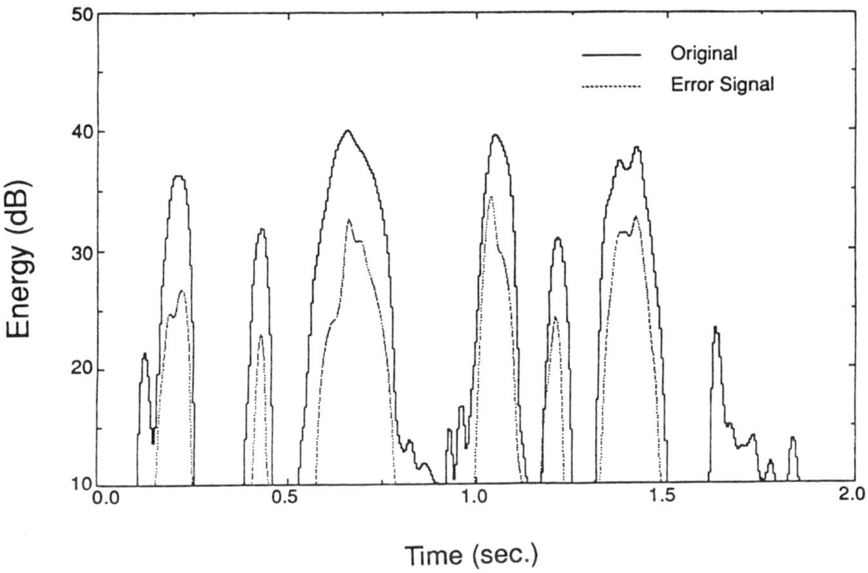

Fig. 4. A comparison of the Energy of the Input Speech Signal and the Error Signal.

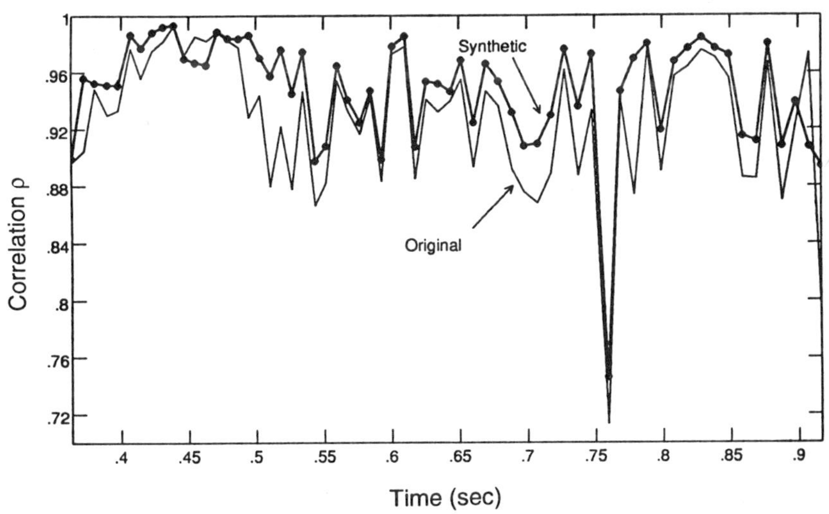

Fig. 5. Correlation Between Successive Pitch-Cycles for the Original and the Reconstructed Speech Signals for a Voiced Segment.

directly to the speech signal, although in such a case one must ensure that the formant peaks are accurately represented. This can easily be achieved by placing certain frequency components in formant regions.

BLOCKWISE INTERPOLATION

As mentioned in the introduction, the parameters of the sparse frequency-domain representation vary smoothly from one pitch-cycle to the next for the slowly evolving segments of voiced speech. In this section, we present a simple scheme to perform blockwise linear interpolation of these parameters over 20-30 ms. Informal listening tests have shown that blockwise interpolation does not introduce any additional audible distortion in the reconstructed speech signal.

During the analysis phase, a local pitch value is determined at the beginning of every frame (for the present discussion, each 20 ms segment of the signal is referred to as a frame). An appropriate pitch-cycle waveform is extracted from the LPC excitation near each frame boundary. The current pitch-cycle waveform is then time-aligned with the pitch-cycle waveform for the previous frame and analyzed using the sparse frequency-domain representation. During the synthesis phase, the pitch and the frequency-domain parameters are interpolated over the current frame. The excitation waveform is then synthesized from the interpolated parameters and passed through the synthesis filter to produce reconstructed speech. A time-domain blockwise interpolation scheme was proposed in [7] using the pitch-cycle waveforms.

Pitch-Cycle Extraction

Let $s(n), n = 0, \ldots, p(t_1)$, represent a pitch-cycle of the speech waveform near the current frame boundary. In the first step, the location of the main feature in a pitch-cycle is determined by maximizing the energy within a small time-window that is shifted within the pitch-cycle. Define the energy $E(i)$ as

$$E(i) = \sum_{l=-L}^{L} \left[w(l)s(i\Delta - l)\right]^2, \qquad i = 0, 1, \ldots, \lceil \frac{p(t_1)}{\Delta} \rceil, \tag{11}$$

where L is the half window length, $p(t_1)$ is the current pitch value, and Δ is the translation step within the pitch-cycle. The window $w(l)$ was selected to be the Bartlett window. The location of the maximum of $E(i)$ provides an initial estimate of the center of the current pitch-cycle waveform.

In the second step, the location of the pitch-cycle waveform is also shifted within the current frame and the correlation with the previous pitch-cycle waveform is computed. The location for which the correlation value is maximum defines the current pitch-cycle waveform. The sparse frequency-domain analysis is then used to obtain the excitation parameters. Only small shift values are required in this step to determine the exact location of the pitch-cycle waveform.

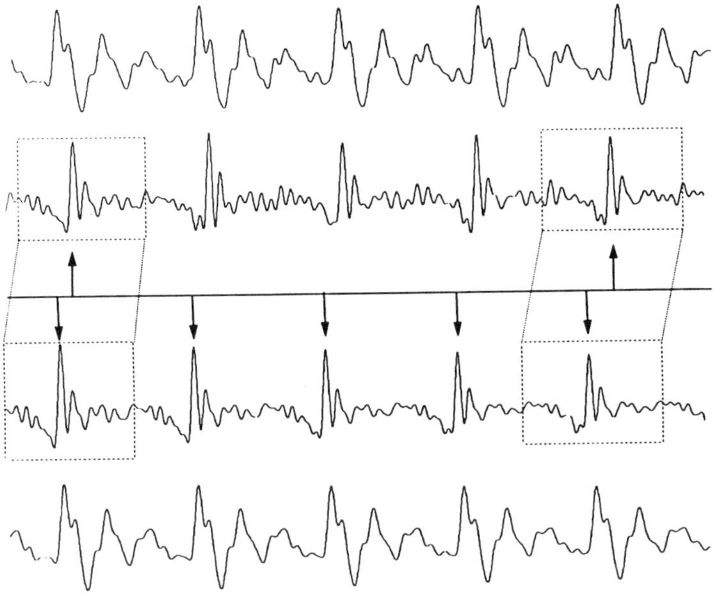

Fig. 6. Blockwise Interpolation: (a) Input Speech, (b) Original LPC Excitation, (c) Location of Extracted Pitch-Cycle Waveforms, (d) Location of Pitch-Cycle Waveforms in Reconstructed Signals, (e) Synthetic LPC Excitation, and (f) Synthetic Speech.

Parameter Interpolation

Let $\{a'_k(t_0), b'_k(t_0)\}$ denote the parameters for the pitch-cycle waveform in the previous frame. Let $\{a'_k(t_1), b'_k(t_1)\}$ denote the excitation parameters for the current frame. Then the parameters $\{a'_k(t), b'_k(t)\}$ at time instant t are given as

$$\begin{aligned} a'_k(t) &= \alpha(t) \cdot a'_k(t_0) + (1 - \alpha(t)) \cdot a'_k(t_1) \\ b'_k(t) &= \alpha(t) \cdot b'_k(t_0) + (1 - \alpha(t)) \cdot b'_k(t_1), \qquad k = 1, \ldots, K, \end{aligned} \tag{12}$$

where $\alpha(t)$ is the interpolation function. For our work, we use a linear interpolation function. The interpolation of pitch is also performed in a similar manner. Due to the linear interpolation of pitch, the number of samples in the current frame of the reconstructed signal may not be equal to the number of samples in the input signal. Hence, the input and the reconstructed speech signals are, in general, not synchronized. This issue is discussed in detail in [4]. Figure 6 shows the result of blockwise interpolation for one frame. The sparse frequency-domain parameters are obtained for the two pitch-cycle waveforms marked by rectangles in Fig. 6(b). The reconstructed excitation is shown in Fig. 6(e).

Informal listening experiments show that there is no additional loss in perceptual quality when blockwise interpolation is used over 20-30 ms segments. Note that for low-pitched male voices, there are relatively few pitch-cycles (2 or 3) within a frame. As a result, interpolation is generally performed between alternate pitch-cycle waveforms. For female voices, interpolation is performed over many pitch-cycles.

CONCLUSIONS

In this paper, we have presented an efficient frequency-domain representation of the LPC excitation for voiced speech that takes advantage of the nonuniform spacing of the critical bands in the auditory system. When the sparse representation is used during voiced segments of speech on a pitch-cycle-by-pitch-cycle basis, only five or six components are sufficient to represent the 1-4 kHz region to obtain synthetic speech without significant loss in perceptual quality. During slowly evolving voiced segments of speech, these parameters are found to vary smoothly from one pitch-cycle to the next. This allows linear interpolation of these parameters over 20-30 ms without introducing any additional audible distortion in the reconstructed speech signal.

Although our experiments used only the voiced segments of each speech utterance, we have found that the sparse frequency-domain representation is also valid for the unvoiced/transitional segments of the utterance. Note that our sparse frequency-domain representation results in smoothing of the spectrum in the high-frequency region. Since the unvoiced/transitional segments have more noise-like characteristics, such smoothing introduces artifacts in the reconstructed speech signal that can be audible. Therefore, for such segments the number of sinusoidal components cannot be made too small. We have found that by using a fixed excitation-block size of 5 ms for unvoiced/transitional frames (20 sinusoidal components for a Fourier series representation), the number of components in our sparse representation can be reduced to 13 without introducing audible distortion in the synthetic speech.

The sparse frequency-domain representation could potentially be used to obtain an efficient speech coding system with reduced bit-rate requirements although the reduction depends strongly upon the quantization characteristics of the parameters.

REFERENCES

[1] Moore, B. C. J. "An Introduction to the Psychology of Hearing," *Academic Press*, pp. 84-136, 1989.

[2] Atal B. S. and Remde J. R., "A New Model for LPC Excitation for Producing Natural Sounding Speech at Low-Bit Rates," *Proc. Int. Conf. on Acoust., Speech, Sig. Proc.*, pp. 614-617, 1982.

[3] Atal B. S. and Schroeder M. R., "Stochastic Coding of Speech Signals at Very Low Bit Rates," *Proc. Int. Conf. Commun. - ICC84*, pp. 1610-1613, 1984.

[4] Kleijn W. B., "Analysis-by-Synthesis Speech Coding Based on Relaxed Waveform-Matching Constraints," Ph.D. thesis, *Delft University of Technology*, The Netherlands, 1991.

[5] Granzow, W. and Atal, B. S., "High-Quality Digital Speech at 4 kb/s," *IEEE Global*

Telecommunications Conference, pp. 941-945, 1991.

[6] Law H. B. and Seymour R. A., "A Reference Distortion System Using Modulated Noise," *Proc. of the Institute of Electrical Engineers*, pp. 484-485, 1962.

[7] Kleijn W. B. and Granzow W., "Methods for Waveform Interpolation in Speech Coding," *Digital Signal Processing 1(4)*, pp. 215-230, 1991.

[8] Mallat S. G., "Multifrequency Channel Decompositions of Images and Wavelet Models," *IEEE Trans., Acoust., Speech, and Sig. Proc.*, vol. 37, No. 12, Dec. 1989.

31

PRODUCT CODE VECTOR QUANTIZATION OF LPC PARAMETERS*

Shihua Wang†, Erdal Paksoy††, and Allen Gersho††

†Teknekron Communication Systems
2121 Allston Way, Berkeley, CA 94704

††Department of Electrical and Computer Engineering
University of California, Santa Barbara, CA 93106

INTRODUCTION

In speech coders based on linear prediction modeling it is important to accurately represent the spectral envelope of each frame to avoid degrading the quality of the synthesized speech. We generally aim for *transparent* quantization of the LPC parameters so that there is no audible difference between coded speech signals synthesized using quantized and unquantized LPC coefficients.

A widely accepted criterion for measuring the accuracy of LPC quantization is the Log Spectral Distortion (SD) measure given by:

$$SD = \left[\frac{1}{2\pi} \int_{-\pi}^{\pi} [\, 10 \log S(\omega) - 10 \log \hat{S}(\omega) \,]^2 d\omega \right]^{1/2}$$

See, for example, [1]. For transparent quantization, it is sufficient to quantize the LPC coefficients with an average SD of up to approximately 1 dB, while holding below 2% the percentage of outlier frames having an SD value greater than 2 dB. With current emphasis on coders operating at 4 kbit/s or below, it has become extremely important to be as efficient as possible in quantizing the LPC parameters while maintaining an adequate quality spectral representation. Although transparent quantization may be a more stringent objective than needed for many low rate coders, this quality benchmark provides a useful reference for comparative assessment of different quantization schemes. Here we report our results for coding of the line spectral frequency (LSF) parameter set with split vector quantization (SVQ), multistage vector quantization (MSVQ), and interframe coding combined with SVQ while maintaining a tolerable level of complexity. Important studies on the same topic have also been made by others, see in particular [2], [3].

*This work was supported in part by the National Science Foundation, Fujitsu Laboratories, Ltd, the UC Micro Program, Rockwell International Corporation, Hughes Aircraft Company, and Eastman Kodak Company.

INTERFRAME LPC QUANTIZATION WITH PRODUCT CODES

In memoryless quantization each input LPC parameter vector is quantized independently of any other input vector. However, for some segments of speech, especially for sustained vowels, the spectral envelope varies slowly and spectra of adjacent frames are highly correlated. This correlation can be handled to some extent by combining a reduced LPC update rate with interpolation of LPC parameters. Alternatively, a quantization method which directly exploits interframe redundancy can efficiently reduce the coding rate for the time varying spectral envelope.

One way to exploit interframe correlation is by vector linear prediction (VLP), introduced for waveform coding in [4] and augmented by switched-adaptive vector prediction. In [5] VLP was applied to switched-adaptive interframe vector prediction (SIVP) of LPC parameters where the prediction error (the error in predicting the current LPC vector from the past), rather than the LPC vector itself, is quantized and transmitted. The line spectral frequencies (LSFs) were chosen as the LPC parameter set. By applying VLP, each spectral parameter in the current frame is predicted not only from the same parameter in prior frames but also from other spectral parameters in prior frames. The potential of VQ in SIVP was also exploited in [5] where the complexity increase was avoided by allocating more bits to the predictor selection and using 12-bit full-search VQ for a rate of 20 bits/frame. Although VQ with SIVP outperformed SQ with SIVP at the same rate, the limited resolution (bits per LPC parameter) that was allocated to VQ still prevented it from attaining transparent quality.

In our study, we combined SIVP with a product-code VQ to encode the LSF spectral parameters with a higher resolution. Since higher than first order prediction was found to deliver little additional gain [5], the LSF vector x_n in the current frame is predicted via the matrix A from the corresponding vector x_{n-1} for the prior frame. Thus, $\hat{x}_n = Ax_{n-1}$ where A is computed from training data. The prediction error vector is $e_n = x_n - \hat{x}_n$. In switched-adaptive vector prediction (SIVP) the matrix A is adapted vector by vector through use of a set of predictors each of which is specifically designed to suit a subset of the input statistical space. Before prediction, each input vector is statistically classified into one of the subsets, the corresponding predictor for that subset is selected, and an index identifying the selected predictor matrix for the frame is transmitted to the receiver.

Split VQ of LSF Prediction Error

For quantizing the prediction error vector we chose the product-code VQ known as split VQ (SVQ) for its desirable properties when used with LSF's. The error vector e is split into two subvectors of dimension 4 and 6 respectively:
$e^1 = \{e(1), e(2), e(3), e(4)\}$ and $e^2 = \{e(5), e(6), e(7), e(8), e(9), e(10)\}$. The two subvectors are then vector quantized separately. The decoder reconstructs the original vector by first decoding each subvector and then concatenating these subvectors to form an approximation of the original vector. SVQ for LSF quantization was first reported briefly in [6], and studied in [1] for single frame LPC coding.

In the encoder a weighted mean squared error (WMSE) criterion is applied to minimize a perceptually meaningful distortion: $WMSE = \sum_{i=1}^{P} w_i [x_i - \hat{x}_i]^2$. Here x_i and $\hat{x}_i$ are the ith component of the P dimensional input and quantized vectors respectively, and w_i is the weighting factor, computed according to the formula given in [7]. The weights are determined by two factors, the spectral sensitivity of different LSF coefficients and the human hearing sensitivity at different frequencies.

Encoder Design and Performance

The encoder has two essential components: a switched-adaptive vector predictor and a split vector quantizer. Since SIVP is actually performed on the quantized LPC parameter vector, these two parts interact with each other: designing the predictor and quantizer separately would not give an optimal performance. Instead, a closed-loop iterative method is used to jointly optimize predictor and quantizer.

Given a training set, the design procedure contains two phases: a) initializing the predictor and quantizer in an open loop and b) iteratively optimizing predictor and quantizer in a closed-loop [4], [5].

In each iteration, all training vectors $\{x(n)\}$ are encoded by the current quantizer and switched-predictor, which are fixed through the entire training iteration. As encoding proceeds, a new training set $\{e(n)\}$ is generated and stored. Meanwhile, the current vector $x(n)$ and the previous reconstructed vector $\tilde{x}(n-1)$ are used to update the new covariance matrices of the selected class for the current input vector. At the end of each iteration, a new quantizer and a new set of prediction matrices are designed from the new training set $\{e(n)\}$ and covariance matrices respectively, and are then used for the next iteration. This process is repeated until the improvement is less than a given threshold.

In the closed-loop optimization we used WMSE as the criterion for quantizer design, and the spectral distortion SD as the performance measure to terminate the iteration.

The speech database on which the encoder design was based contains 26 minutes of speech, recorded with different acoustic backgrounds and a variety of speakers. The training set contained 42235 LSF vectors, each derived from a 30 ms speech frame sampled at 8 kHz. Three sequences each one minute long, coming from totally different acoustic environments, were used to test the robustness of the encoder outside the training set. The performance of the encoder was objectively assessed via the average SD and the percentage of outliers (vectors with SD > 2 dB).

To explore the rate trade-off between predictors and quantizers, we compared different bit allocations for rates of 24 and 25 bits/frame. The results are summarized in Table 1. The SIVP column gives the number of bits to specify the predictor in each frame, however, zero bits for SIVP means that no predictor is used, i.e., direct band-splitting VQ is used without any interframe coding.

The results in Table 1 suggest the following conclusions. (i) When SIVP is applied, VQ resolution dominates the encoder performance. (ii) Allocating more bits to SIVP degrades performance only slightly, but saves much complexity, a fact that

Total bits	Bit allocation			Inside TS		Outside TS	
	SIVP	VQ1	VQ2	Avg. SD (dB)	Outliers >2 dB(%)	Avg. SD (dB)	Outliers >2 dB(%)
24	0	12	12	1.04	2.37	1.14	1.74
24	1	11	12	0.96	2.61	1.01	2.39
24	1	12	11	1.03	3.38	1.02	2.55
24	2	11	11	1.05	3.89	1.05	2.73
24	3	10	11	1.06	4.02	1.07	2.66
24	3	11	10	1.11	5.05	1.08	2.94
24	4	10	10	1.09	4.68	1.13	2.88
25	1	12	12	0.92	1.88	0.96	1.91
25	2	11	12	0.95	2.43	1.01	1.92
25	4	10	11	1.00	3.09	1.01	2.01

Table 1: Comparison of different bit allocations for coding rate of 24 and 25 bits/frame.

may be exploited in a real-time implementation, and (iii) The mean values of SD for encoders with SIVP at high VQ resolution (i.e., 12 bits for each stage) are lower than for direct VQ, but this is offset by a somewhat higher percentage of outliers for the results outside the training set.

Table 2 shows various bit rates and best allocations in the range 22 – 25 bits/frame, together with the attainable performance.

Bit allocation				Inside TS		Outside TS	
Total	SIVP	VQ1	VQ2	Avg. SD (dB)	Outliers >2 dB(%)	Avg. SD (dB)	Outliers >2 dB(%)
22	1	10	11	1.12	5.61	1.18	3.50
23	1	11	11	1.07	4.17	1.09	2.95
24	1	11	12	0.96	2.61	1.01	2.39
25	1	12	12	0.92	1.88	0.96	1.91

Table 2: Performance of VQ with SIVP at different coding rates.

We compared our 24 bit per frame LPC encoding result at a rate of 800 bit/s in Table 2 with two other low rate LPC coding results at 24 bits per frame. Neither of these uses interframe coding. In [1] an average SD of 1.03 dB is reported with 1.03% outliers and a bit rate of 1200 bit/s. In [3] an average SD of 1.14 dB with 1.40% outliers and a bit rate of 1066 bit/s is reported. Thus, our method has a lower coding rate and a smaller average spectral distortion, but a higher percentage of outliers. Our outlier percentage may be related to differences in sentence lengths in the databases used for test and design (3 – 4 seconds in ours versus an average of 7 seconds in [1]) since short sentences place a heavy burden on SIVP.

VECTOR QUANTIZATION OF LSF PARAMETERS USING GENERALIZED PRODUCT CODES

We now examine an LSF quantization method that does not make use of interframe coding but aims to achieve very high coding efficiency through a more extensive use of the product code concept. In product code VQ, the feature vectors are extracted from the input vector and sequentially quantized using feature codebooks. The quantized features are recombined using a synthesis function to reconstruct the quantized version of the original vector. We have studied the quantization of the LSFs using both SVQ and MSVQ. In SVQ, a vector is partitioned into subvectors each of smaller dimensionality which are quantized separately. The quantized subvectors are then concatenated to form a quantized vector. In MSVQ the entire input vector is first quantized with a first stage codebook. The error vector between the input vector and the output of the first stage quantizer is then quantized using a second stage codebook. This process is repeated for the remaining stages. The quantized input vector is then obtained by adding together the outputs of all the stages.

In the conventional product codes each feature is quantized using one codebook. Within the framework of generalized product codes [8], [9], it is possible to use more than one codebook to quantize a given feature. The number of codebooks used to quantize a particular feature is called the *fanout* for that feature. These codebooks can be designed one stage at a time using the constrained storage vector quantization (CSVQ) algorithm described in [10]. A more detailed description of generalized product codes can be found in another chapter in this volume [11].

Improvement of Performance using Multiple Survivors Search

For most product codes, sequential search of the feature codebooks usually yields suboptimal performance with respect to the product codebook. A *multiple survivor* search method can improve performance by selecting the L_i codevectors with the L_i lowest distortion values for each feature f_i. In the final stage, the best candidate out of the final survivor set is selected to minimize the overall distortion for the full input vector. The multiple survivor concept was described in [9] and a similar approach for encoding LSFs was reported in [2]. It should be noted that the search complexity for MSVQ encoding may increase substantially with the use of multiple survivors. However, with two-stage MSVQ, the search complexity increases only linearly with the number of first stage survivors.

Distortion Measures for Design and Encoding

Our main performance objective to be minimized is the SD between quantized and unquantized LPC parameter vectors. However, due to the complex dependence of LSFs and the spectral envelope involved in SD computation there is no easy way of incorporating the SD into the codebook design algorithm. For this reason WMSE is used to design the codebooks. Here the weights are computed for each LSF vector using the formulas in [1]. While the SD is difficult to use in codebook design, it can be used in encoding. However, because of the computational cost of calculating SD, it is not realistic to perform the entire codebook search using that distortion function. On the other hand when a multiple survivor method is employed, SD can be used to select the best candidate from the final stage survivor set, (see [2]). This was done in the case of SVQ and MSVQ, where L_i survivors were chosen from each feature codebook and the SD was used to select the best quantized output from the overall survivor set.

LPC Analysis

In order to obtain a training set of LPC vectors, LPC analysis was performed on a large speech database. The speech was lowpass filtered at 3.4 kHz and sampled at 8 kHz. We performed 10th order LPC analysis using the modified covariance method with high frequency compensation. The analysis window size was 20 ms, and the LPC vector training set contained 144800 vectors. Bandwidth expansion was also used, i.e. we multiplied each prediction coefficient by γ^i, where $i=1, \cdots, 10$ and γ is a constant equal to 0.996. The performance of the quantizers was evaluated using a test file, 7700 frames (2.5 min) long, independent from the training set.

Results

Figure 2 shows the variation of SD as a function of the bit rate for SVQ designed using WMSE under different fanout conditions. The following conclusions can be drawn from this figure. In SVQ, selecting the best candidate out of a set of 8 survivors gives a savings of one bit/frame, i.e. we get the same average SD value for 8 candidates and 23 bits/frame as we get for 24 bits/frame without multiple survivors. Using four second stage codebooks for SVQ also saves about one bit/frame. Hence, when a fanout greater than one and a multiple survivor search method are used, it is possible achieve the same performance as the quantizer reported in [1], while using 22 bits/frame, thereby saving 2 bits/frame. The proportion of outlier frames having a spectral distortion value above 2 dB is also held under 2% for all quantizers having an average SD value below 1.15 dB.

Finally Table 3 shows the performance of weighted MSVQ. MSVQ has an advantage of 1 bit/frame over SVQ. This is to be expected since SVQ can be viewed as a special case of MSVQ, where some vector components are set to zero. We also observe from Table 3 that using a multiple survivors search where the decision between 8 candidates is based on SD leads to a savings of 2 bits per frame. Hence this method allows us to transparently quantize the LPC vectors at 21 bits/frame. We also noticed that a fanout of four does not yield a tangible improvement in the SD performance of MSVQ. This can be explained by the fact that in the case of MSVQ the second feature statistics do not benefit from the ordering property of the LSFs, and hence have more homogeneous statistics compared to those of the second feature in SVQ. The use of multiple survivors increases the search complexity, nevertheless, the overall complexity of each case considered in Table 3 remains within a few percent of the computational capacity of current digital signal processor chips. A more extensive study of LPC quantization with generalized product codes

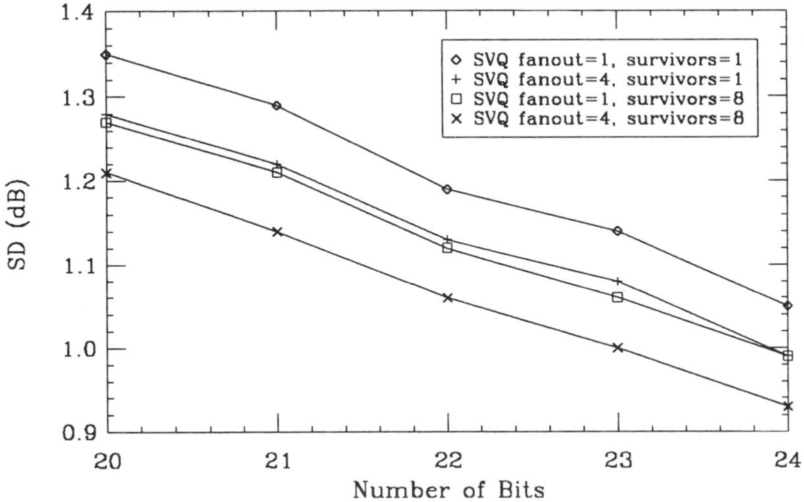

Fig. 2: Performance of Split Vector Quantization of LSF using Generalized Product Codes.

Bit Allocation			Fanout	Survivors	Avg SD	Outliers
Total	Feature 1	Feature 2			(dB)	>2 dB(%)
24	12	12	1	1	1.00	1.35
23	12	11	1	1	1.07	1.87
22	11	11	1	1	1.14	2.87
21	11	10	1	1	1.21	4.19
20	10	10	1	1	1.29	6.30
24	12	12	1	8	0.88	0.35
23	12	11	1	8	0.94	0.62
22	11	11	1	8	1.00	0.74
21	11	10	1	8	1.07	1.71
20	10	10	1	8	1.14	2.21

Table 3: Performance of MSVQ using WMSE.

will be reported in [12].

References

[1] K. K. Paliwal and B. S. Atal, ''Efficient Vector Quantization of LPC Parameters at 24 Bits/Frame,'' *Proc. IEEE Int. Conf. Acoust., Speech, Sign. Processing*, pp. 661-664, Toronto, Canada, May, 1991.

[2] B. Bhattacharya, W. LeBlanc, S. Mahmoud, and V. Cuperman, "Tree Searched Multi-stage Vector Quantization of LPC Parameters For 4 Kb/s Speech Coding," *Proc. IEEE Int. Conf. Acoust., Speech, Sign. Processing*, vol. 1, pp. 105-108, San Francisco, March, 1992.

[3] R. Laroia , N. Phamdo, and N. Farvardin, "Robust and Efficient Quantization of Speech LSF Parameters Using Structured Vector Quantizers," *Proc. IEEE Int. Conf. Acoust., Speech, Sign. Processing*, pp. 641-644, Toronto, Canada, May, 1991.

[4] V. Cuperman and A. Gersho, "Vector Predictive Coding of Speech at 16 kbits/s," *IEEE Transactions on Communications*, vol. COM-33, pp. 685-696, July 1985.

[5] M. Yong, G. Davidson, and A. Gersho, "Encoding of LPC Spectral Parameters Using Switched-Adaptive Interframe Vector Prediction," *Proc. IEEE Int. Conf. Acoust., Speech, Sign. Processing*, vol. 1, pp. 402-405, New York City, April 1988.

[6] S. Wang and A. Gersho, "Phonetically-Based vector excitation coding of speech at 3.6 kbps," *Proc. IEEE Int. Conf. Acoust., Speech, Sign. Processing*, pp. 49-52, Glasgow, May 1989.

[7] G. S. Kang and L. J. Fransen, "Application of Line-Spectrum Pairs to Low-Bit-Rate Speech Encoders," *Proc. IEEE Int. Conf. Acoust.,Speech, Sign. Processing*, pp. 244-247, Tampa, March 1985.

[8] W.-Y. Chan and A. Gersho, "Enhanced Multistage Vector Quantization with Constrained Storage," *Proc. 24th Asilomar Conf. Circuits, Systems, and Computers*, pp. 659-663, Pacific Grove, CA, November 1990.

[9] W.-Y. Chan, "The Design Of Generalized Product-Code Vector Quantizers," *Proc. IEEE Int. Conf. Acoust., Speech, Sign. Processing*, , vol. 3, pp. 389-392, San Francisco, March, 1992.

[10] W.-Y. Chan and A. Gersho, "Constrained-Storage Quantization of Multiple Vector Sources by Codebook Sharing," *IEEE Transactions on Communications*, vol. 39, pp. 11-13, January 1991.

[11] W.-Y. Chan and A. Gersho, "High Fidelity Audio Coding with Generalized Product Code VQ," *Speech and Audio Coding For Wireless and Network Applications (B. Atal, V. Cuperman, A. Gersho editors)*, Kluwer Academic Publishers, (this volume), 1993.

[12] E. Paksoy, W.-Y. Chan , and A. Gersho, "Vector quantization of speech LSF parameters with generalized product codes," *Proc. Int. Conf. Spoken Language Processing*, Banff, Canada, October 1992.

32

A MIXED EXCITATION LPC VOCODER WITH FREQUENCY-DEPENDENT VOICING STRENGTH

Alan V. McCree and Thomas P. Barnwell III

School of Electrical Engineering
Georgia Institute of Technology
Atlanta, GA 30332

INTRODUCTION

Traditional pitch-excited LPC vocoders use a fully parametric model to efficiently encode the important information in human speech. These vocoders can produce intelligible speech at low data rates (800-2400 bps), but they often sound synthetic and generate annoying artifacts such as buzzes, thumps, and tonal noises. These problems increase dramatically if acoustic background noise is present at the speech input. This paper presents a new LPC vocoder model which preserves the low bit rate of a fully parametric model, but adds more free parameters to the excitation signal so the synthesizer can mimic more characteristics of natural human speech. The new model also eliminates the traditional requirement for a binary voicing decision, so the vocoder performs well even in the presence of acoustic background noise.

The new LPC model is based on the traditional LPC vocoder with either a periodic impulse train or white noise exciting an all-pole filter, but contains four additional features as shown in Figure 1. The synthesizer has the following added capabilities: mixed pulse and noise excitation, periodic or aperiodic pulses, pulse dispersion filter, and adaptive spectral enhancement. The following section describes the new model in more detail, including the purpose and design approach for each of these features. The final section of the paper discusses the implementation and evaluation of a 2400 bps LPC vocoder based on this model.

THE NEW LPC VOCODER MODEL

Mixture Excitation

The most annoying aspect of LPC vocoder speech output is a strong buzzy quality. This problem seems to come from the inability of a simple pulse train to reproduce all kinds of voiced speech, so vocoders have previously been proposed with mixtures of pulse and noise excitation [1, 2]. Mixture excitations are commonly used in formant synthesizers [3, 4], and have also been applied in the context of sinusoidal coding [5, 6]. The mixed excitation structure developed for this coder [7, 8] can generate an excitation signal with different mixtures of pulse and noise in each of a large number (4-10) of frequency bands. As shown in Figure 1, the pulse train and noise sequence are each passed through time-varying frequency shaping filters

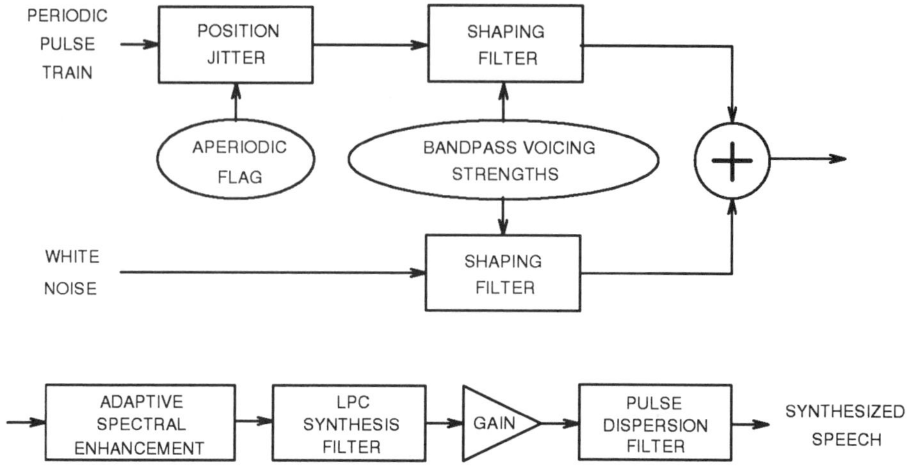

Figure 1: New LPC Synthesizer

and then added together to give a fullband excitation. For each frame, the frequency shaping filter coefficients are generated by a weighted sum of fixed bandpass filters. The pulse filter is calculated as the sum of each of the bandpass filters weighted by the voicing strength in that band. The noise filter is generated by a similar weighted sum, with weights set to keep the total pulse and noise power constant in each frequency band. These two frequency shaping filters combine to give a spectrally flat excitation signal with a staircase approximation to any desired noise spectrum.

For ideal bandpass filters, the excitation signal generated by this approach will have a flat power spectrum as long as the sum of the pulse and noise power in each frequency band is kept constant. The important parameters in a practical filter design are the passband and stopband ripple and the amount of pulse distortion. We implement the filter bank with FIR filters designed by windowing the ideal bandpass filter impulse responses with a Hamming window. This design technique yields linear phase FIR filters with good frequency response characteristics and the additional benefit of a nice reconstruction property: the sum of all the bandpass filter responses is a digital impulse. Therefore, if all bands are fully voiced, the fullband excitation will be an undistorted pulse. Figure 2 shows the frequency responses of a nonuniform five band design.

To make full use of this mixed excitation synthesizer, we need to accurately estimate the degree of voicing in each frequency band. We have developed an algorithm which combines two methods of analysis of the bandpass filtered input speech. First, the periodicity in each band is estimated using the strength of normalized autocorrelations around the pitch lag. This technique works well for stationary speech, but the correlation values can be too low in regions of varying pitch. The problem is worst at high frequencies, and results in a slightly whispered quality to the synthetic speech. The second method uses a technique similar to time domain analysis of the wideband

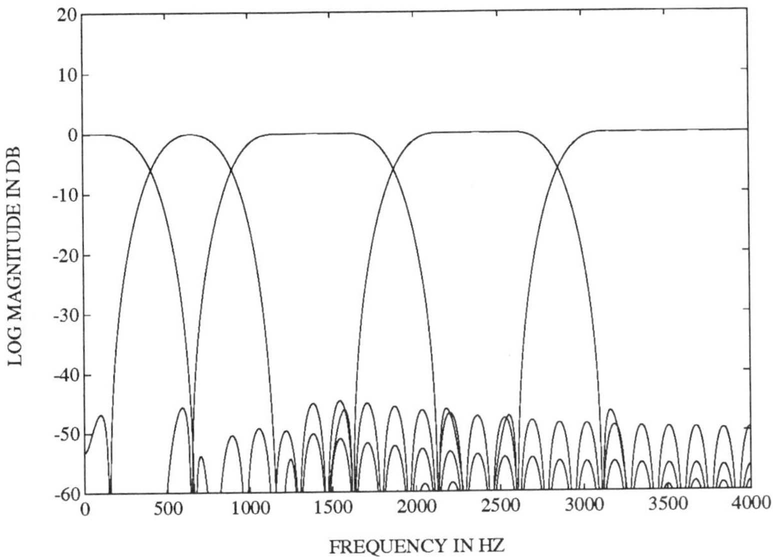

Figure 2: Example Bandpass Filter Responses, 5 band, 48th order

spectrogram to estimate the voicing strength. The envelopes of the bandpass filtered speech are generated by full wave rectification and lowpass filtering, with a notch filter to remove the DC term from the output. At higher frequencies, these envelopes can be seen to rise and fall with each pitch pulse, just as in the wideband spectrogram. Autocorrelation analysis of the envelopes yields an estimate of the amount of pitch periodicity. Since the peaks in this envelope signal are quite broad, small pitch fluctuations have little effect on the correlation values.

Aperiodic Pulses

This mixed excitation can remove the buzzy quality from the LPC speech output, but another distortion is sometimes apparent. This is the presence of short isolated tones in the synthesized speech, especially for female speakers. The tones can be eliminated by varying each pitch period length with a random jitter, but this introduces a hoarse quality in strongly voiced speech segments. Therefore, we have added a third voicing state to the voicing decision which is made at the transmitter [9]. The input speech is now classified as either voiced, jittery voiced, or unvoiced. In both voiced states, the synthesizer uses a pulse/noise mixed excitation, but in the jittery voiced state an aperiodic pulse train is used. Strong voicing is defined by a high correlation in the pitch search at the transmitter, and jittery voicing is defined by either marginal correlation or peakiness in the input signal. The carefully controlled use of aperiodic pulses can remove the tonal noises without introducing additional distortion. It is interesting to note that using aperiodic pulses without mixed excitation does not reduce the buzz, so we presume that the buzzy quality comes from excessive peakiness in the higher frequency bands, while excess periodicity causes tonal noises.

Waveform Matching

By combining mixed excitation with aperiodic pulses, the new LPC vocoder largely avoids major artifacts such as buzz, thumps, and tonal noises. However, the synthetic speech still has a slightly unnatural quality. In comparing bandpass filtered envelopes of input and processed speech, we have noticed some differences in waveforms. Sometimes both waveforms are clearly voiced, but the LPC speech has a more pronounced difference between peak and valley levels. At frequencies near the formants, this could be due to improper LPC pole bandwidth. The synthetic time signal may decay too quickly because the LPC pole has a weaker resonance than the true formant. At frequencies away from the formants, the synthetic excitation signal may have a peak which is too sharp. In natural speech, the excitation may not all be concentrated at the point in time corresponding to glottal closure. This could be due to a secondary peak from the opening of the glottis, incomplete glottal closure, or a small amount of acoustic background noise.

The new LPC vocoder model has two features to remove these problems. To help match the formant resonances, adaptive spectral enhancement is applied with a pole/zero filter based on the LPC coefficients [10]. This boosts the frequencies around the formants in the synthetic speech, and it also provides a better waveform match to natural bandpass filtered speech. In addition, a fixed pulse dispersion filter based on a spectrally flattened synthetic glottal pulse is used. The filter coefficients are based on a triangle pulse which is spectrally flattened using a Fourier series expansion [7]. This filter introduces time-domain spread into the synthetic speech in order to more closely match natural speech waveforms in frequency bands away from the formants.

IMPLEMENTATION AND EVALUATION

In order to evaluate the performance of this new LPC vocoder model, we have implemented a 2400 bps mixed excitation LPC vocoder on a personal computer using the C language with some in-line assembler instructions. The system runs in real-time using two plug-in boards based on the TMS320C30 DSP chip. The bit allocation for this coder is shown in Table 1. The LPC coefficients are determined by the autocorrelation technique using a Hamming window of length 25 msec and coded using scalar quantization of the Line Spectrum Pair (LSP) parameters. The pitch is estimated from a search of normalized correlation coefficients of the lowpass filtered LPC residual signal, with an explicit check for possible pitch doubling. To improve performance in noise, a second search is performed on the lowpass filtered input speech and gross pitch errors are corrected based on one past and one future frame. The overall voicing decision is based primarily on the strength of the pitch periodicity. Strong pitch correlation results in classification as strongly voiced, and a frame with marginal pitch correlation or high peakiness in the residual signal is classified as jittery voiced so that aperiodic pulses will be used. An unvoiced frame is declared if none of

LPC coefficients (10 LSP's)	34
gain (2 per frame)	8
pitch and overall voicing	7
bandpass voicing	5-1
aperiodic flag	1
TOTAL: 54 bits / 22.5 msec = 2400 bps	

Table 1: 2400 bps LPC Vocoder Bit Allocation

these conditions are met. The mixed excitation in the lowest frequency band is based on the overall voicing state, while the higher four bands each have their own binary voicing decision.

The new 2400 bps LPC vocoder has undergone both informal and formal listening tests. The coder has been compared to two standard speech coders: 2400 bps DoD LPC-10e v.55 and 4800 bps DoD CELP release 3.2 [11]. Informal listening on a database of about 20 speakers shows that the new coder generates high quality speech which approaches the performance of the higher bit rate CELP coder. Both male and female speakers are accurately reproduced. In addition, the new coder maintains good performance in acoustic background noise, unlike the DoD LPC-10e. In a synthetic white noise environment, the mixed excitation produces natural sounding speech without obvious artifacts such as buzz or thumps. In standard military communications environments such as airplanes, tanks, and helicopters the new coder still produces natural sounding speech, although the noise itself sounds somewhat distorted.

Formal Diagnostic Acceptability Measure (DAM) testing has been performed on the new 2400 bps LPC vocoder and the two standard speech coders. All the coders were simulated on a Sun workstation. The tests were run on a speech database consisting of twelve sentences from each of three male speakers and three female speakers. Additional testing was done with synthetic white noise added to the same speech input. The noise was generated by a Gaussian random number generator, and the signal to noise ratio over the six speaker database was about 8 db. The DAM test results for both clean and noisy speech are shown in Table 2. The clean speech DAM

Speech Coder	Clean	Noisy
2400 bps DoD LPC-10e	54.0	33.9
2400 bps new LPC	58.9	41.0
4800 bps DoD CELP	62.6	39.4

Table 2: 6 Speaker DAM Test Scores for Clean and Noisy Inputs

scores show that the 2400 bps mixed excitation LPC vocoder produces speech which is close in quality to the 4800 bps DoD CELP. For the noisy speech, all the scores are low due to the annoying amount of background noise, but the speech can still be clearly understood. In this difficult environment, the new coder performs better than the higher rate standard.

REFERENCES

[1] J. Makhoul, R. Viswanathan, R. Schwartz, and A. W. F. Huggins, "A Mixed-Source Model for Speech Compression and Synthesis," *J. Acoust. Soc. Amer.*, vol. 64, pp. 1577–1581, Dec 1978.

[2] S. Y. Kwon and A. J. Goldberg, "An Enhanced LPC Vocoder with no Voiced/Unvoiced Switch," *IEEE Trans. Acoust., Speech, Signal Processing*, vol. 32, pp. 851–858, Aug 1984.

[3] D. H. Klatt, "Review of Text-to-speech Conversion for English," *J. Acoust. Soc. Amer.*, vol. 82, pp. 737–793, Sep 1987.

[4] J. N. Holmes, "The Influence of Glottal Waveform on the Naturalness of Speech from a Parallel Formant Synthesizer," *IEEE Trans. Audio and Electroacoustics*, vol. 21, pp. 298–305, June 1973.

[5] D. W. Griffin and J. S. Lim, "Multiband Excitation Vocoder," *IEEE Trans. Acoust., Speech, Signal Processing*, vol. 36, pp. 1223–1235, Aug 1988.

[6] R. McAulay, T. Parks, T. Quatieri, and M. Sabin, "Sine-Wave Amplitude Coding at Low Data Rates," in *Advances in Speech Coding*, pp. 203–214, Norwell MA: Kluwer Academic Publishers, 1991.

[7] A. V. McCree and T. P. Barnwell III, "Improving the Performance of a Mixed Excitation LPC Vocoder in Acoustic Noise," in *Proc. IEEE Int. Conf. Acoust., Speech, Signal Processing*, pp. II137–II140, 1992.

[8] A. V. McCree, *A New LPC Vocoder Model for Low Bit Rate Speech Coding*. PhD thesis, Georgia Institute of Technology, August 1992.

[9] A. V. McCree and T. P. Barnwell III, "A New Mixed Excitation LPC Vocoder," in *Proc. IEEE Int. Conf. Acoust., Speech, Signal Processing*, pp. 593–596, 1991.

[10] J. H. Chen and A. Gersho, "Real-Time Vector APC Speech Coding at 4800 bps with Adaptive Postfiltering," in *Proc. IEEE Int. Conf. Acoust., Speech, Signal Processing*, pp. 2185–2188, 1987.

[11] J. P. Campbell Jr., T. E. Tremain, and V. C. Welch, "The DoD 4.8 kbps Standard (Proposed Federal Standard 1016)," in *Advances in Speech Coding*, pp. 121–133, Norwell MA: Kluwer Academic Publishers, 1991.

33

ADAPTIVE PREDICTIVE CODING WITH TRANSFORM DOMAIN QUANTIZATION

Udaya Bhaskar

COMSAT Laboratories
22300 COMSAT Drive
Clarksburg, MD 20871,USA

INTRODUCTION

Toll quality encoding of voice at 16 kbit/s has been an area of active research and development for a number of years. However, as evidenced by a recent study conducted by COMSAT Laboratories, few techniques have proven capable of meeting the rigorous requirements for toll quality. Existence of toll quality 16 kbit/s coders with robustness to channel impairments and cost-effective implementation can spur the growth of a number of newly emerging technologies such as Personal Communications and Low Carrier to Noise Ratio (C/N) Satellite Systems.

This paper presents the Adaptive Predictive Coding with Transform Domain Quantization (APC-TQ), a voice coding technique that has demonstrated toll quality voice performance at 16 kbit/s rate. This was established by formal subjective testing covering an extensive range of conditions, including eight talkers, three languages and three bit error conditions. The APC-TQ achieved a voice performance that was better than that of the CCITT Standard G.721 32 kbit/s coder under all conditions tested.

In addition, APC-TQ has other advantages. The encoded bits can be prioritized and selective error protection can be incorporated, resulting in robust operation over marginal links. The transmission rate is also easily variable, with a graceful change in voice quality. The scalar quantization version of APC-TQ is computationally simple and a full duplex codec has been implemented on a single fixed point digital signal processor.

APC-TQ CODING TECHNIQUE

APC-TQ is a block adaptive predictive coding technique. The input signal is processed in frames of 128 samples. Signal redundancies are removed by short term and long term predictors. A tenth order Burg analysis method is used to determine the short term predictor. The long term predictor parameters are selected from a 7 bit vector quantizer codebook to minimize the total squared residual error.

APC-TQ differs from other predictive coders in the technique used for quantization of the residual. The conventional approach is to quantize the time domain residual, within a quantization noise feedback loop, which controls the power spectrum of the reconstruction noise. This approach is susceptible to instabilities when processing signals with large spectral dynamic range, such as resonant voice sounds and sinusoids.

TRANSFORM DOMAIN QUANTIZATION

In the APC-TQ technique, the transform coefficients obtained by an orthogonal transformation of the residual are quantized. The power spectrum of the reconstruction noise is controlled by a nonuniform distribution of the available bits, in the case of scalar quantization, or by a spectral weighting function in the case of vector quantization.

Transform quantization differs in important respects from transform coding, in which the function of the transform, applied to the input signal, is diagonalization of the covariance matrix of the transform coefficients. In contrast, transform quantization is applied to the residual, a highly uncorrelated signal which will yield transform coefficients with a diagonal covariance matrix for any orthogonal transform. This creates the interesting possibility of using transforms which have so far not been used for voice signals.

Secondly, for voice signals, more accurate signal decorrelation is achieved by forward adaptive linear prediction than by a suboptimal transform such as the discrete cosine transform (DCT). Since the predictors are optimized for each frame, the spectral dynamic range of the input signal can be translated to higher coding gain by prediction based approaches such as APC-TQ, than by transform coders.

Quantization of Transform Coefficients

Two approaches have been developed: a scalar quantization approach which is computationally simpler and a vector quantization approach that can achieve higher performance.

Scalar Quantization

In the scalar quantization approach, 256 bits are distributed nonuniformly among the 128 transform coefficients. The distribution is determined by the desired reconstruction noise power spectrum and is based on the input signal power spectrum, as estimated by the short and long term prediction parameters. Bit allocation to each transform coefficient is in proportion to the input signal power (in dB) at the corresponding frequency. Each transform coefficient is allocated between 0 and 5 bits, and is scalar quantized by an optimized Max quantizer. A block diagram of the APC-TQ encoder with scalar quantization is shown in Figure 1.

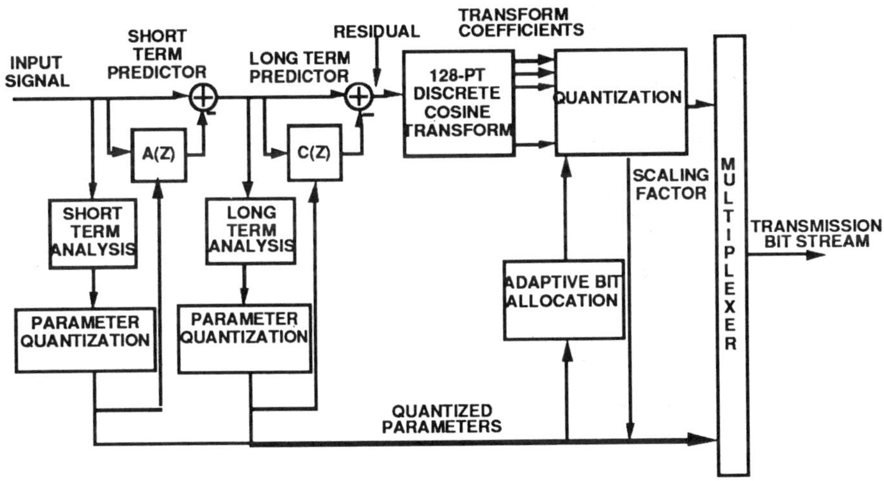

Figure 1. Block Diagram of APC-TQ Encoder with Scalar Quantization

Vector Quantization

In the vector quantization approach, the transform coefficients are grouped into 16 vectors of dimension 8. Vector formation is dependent upon the input signal power spectrum, as estimated by the short and long term prediction parameters. Each vector is formed such that the total input signal power (in dB) at the frequencies corresponding to the transform coefficients of the vector, is uniform among all vectors. Each vector is quantized by exhaustive search in a 12 bit random codebook, with a weighted squared error distortion measure. The weighting of the error vector controls the reconstruction noise power spectrum. The weighting is determined by the short and long term predictor parameters as well as the desired noise spectral shaping.

DECODER

Since the bit allocation in the scalar quantization approach and the adaptive vector formation in the vector quantization approach are both backward adaptive, the transform coefficients can be properly decoded, in the absence of bit errors. Inverse transformation is performed and the resulting signal is used as an excitation to the cascade of long term and short term synthesis filters to generate an approximation to the input signal. Figure 2 shows a block diagram of the APC-TQ decoder for the case of scalar quantization.

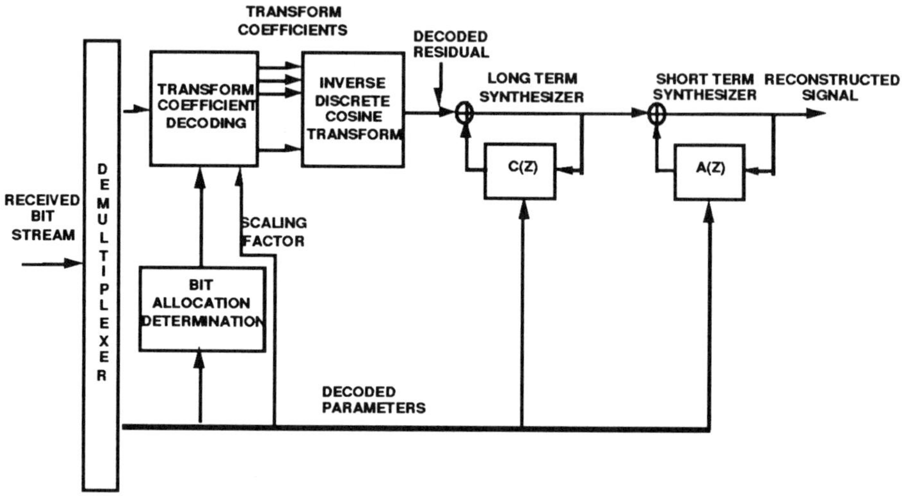

Figure 2. Block Diagram of APC-TQ Decoder with Scalar Quantization

PERFORMANCE

To date, APC-TQ with DCT and scalar quantization has been implemented and tested extensively. Under all conditions tested, the subjective performance of the coder was better than that of the CCITT G.721 32 kbit/s ADPCM coder and only marginally worse than that of the CCITT G.711 64 kbit/s PCM coder. Figure 3 shows the results obtained from English language subjective tests.

With certain modifications, satisfactory performance has been obtained for CCITT System #5 Interregister Signalling and DTMF signalling as well as for voiceband data at rates of 300 bit/s to 2400 bit/s (i.e., V.21, V.22, V.22bis, V.23 and V.27ter). The characteristics of the APC-TQ coder have also been assessed against the requirements specified in CCITT Recommendation G.712 (eg. amplitude and group delay distortion, signal to quantization noise ratio).

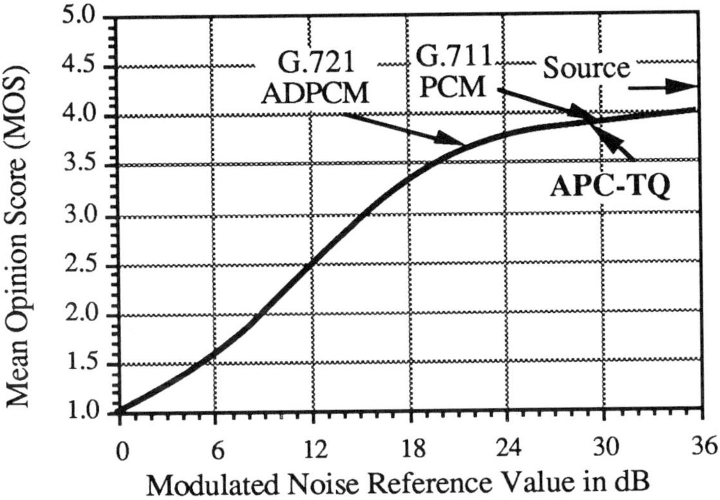

Figure 3. English Subjective Test Results

34

FINITE-STATE VQ EXCITATIONS FOR CELP CODERS

Adil Benyassine, Dept. of ECE, NJIT, Newark, NJ 07102

Hüseyin Abut and Gonçalo C. Marques, ECE Department, San Diego State University, San Diego, CA 92182.

INTRODUCTION

Important advances in speech compression algorithms and the availability of efficient low cost signal processors to implement these algorithms resulted in systems which can reproduce reasonably good quality speech at bit rates as low as 4,800 bits per second. The success of these coders has stimulated considerable interest both in the scientific research community and in industrial development centers in using the low bit rate speech coding technology for emerging real-life applications. These include the cellular telephone services, the secure telephone, mobile satellite, land mobile communications, and the multi-media applications.

Code Excited Linear Predictive (CELP) coders and their derivatives[1] use either an ensemble of pulses or a vector of signal prototypes as the excitation models. The performance of these coders at present degrades rapidly below 4,800 b/s. The rapid growth in demand for cellular communications at half rate (4,000 b/s) has made it necessary to have a careful look at various components of these coders. Important progress has been made in recent years in reducing the bit rate for encoding the LPC parameters[2]. However, the percentage of the bit rate required for encoding the input excitation continues to remain high. Here we propose to improve the excitation model of CELP coders by *embedding* a variation of a Finite-State Vector Quantizer (FSVQ)[3] to structure the excitation process for reducing the bit rate with a "graceful" degradation in synthesized speech quality.

FSVQ CELP CODER STRUCTURE

The basic structure of the FSVQ CELP coder is depicted in Figure 1.The fundamental differences between our system and the basic CELP coding architecture are the inclusion of a *finite-state classifier* in the loop and the way a 40 samples (5.0 ms) long vector is composed from a number of codebooks. We have included a *"derailing algorithm"* to guarantee a regular resetting of the finite-state machine at no extra cost. In addition, we have generated codebooks using variations of the *greedy tree growing algo-*

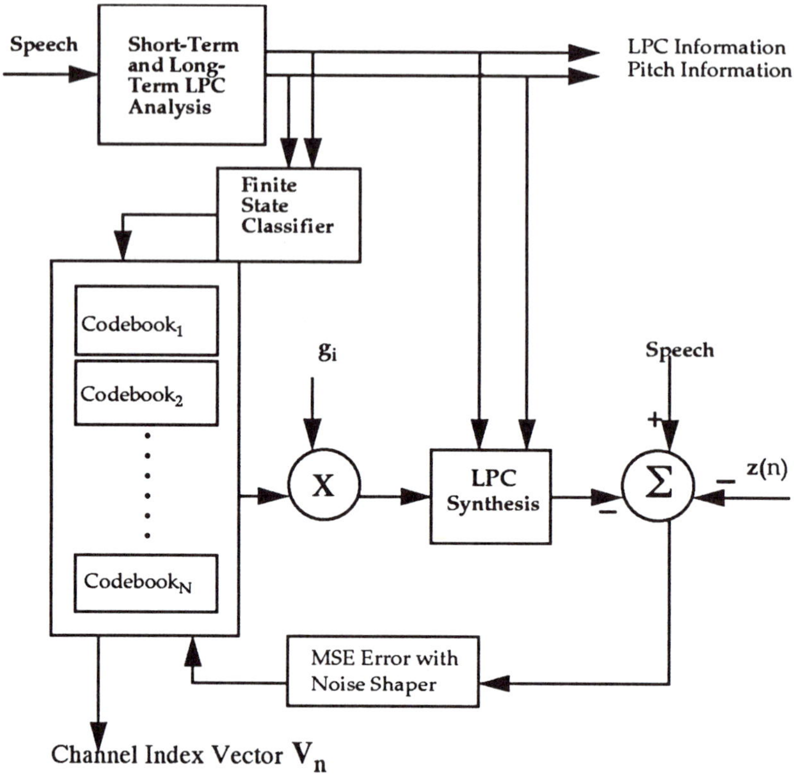

Figure 1. Block Diagram of the FSVQ-CELP Coder.

rithm of Riskin and Gray [5]. The remaining blocks in the coder are identical to those of other similar CELP coders.

Analysis: The short-term linear prediction analysis is performed on an open-loop fashion once every 20 ms using a preemphasis factor of 0.95, a Hamming window, a bandwidth expansion factor of 15 Hz and an autocorrelation analysis of order 10. The analysis window is centered at the end of the last pitch frame. A memoryless vector quantizer was designed for these coefficients using the basic VQ design algorithm subject to the Itakura-Saito spectral distortion measure [4]. Our long-term analysis was based on an open-loop second-order pitch prediction with quarter sample uniform resolution over 5 ms pitch windows. The transfer function P(z) of a second-order pitch predictor with delay of M samples and predictor coefficients $\{\beta_1, \beta_2\}$ is given by

$$P(z) = 1 - \beta_1 z^{-M} - \beta_2 z^{-(M+1)} \qquad (1)$$

The quarter sample resolution was achieved via a simple linear interpolation between two prediction coefficients. The search was a full search in the pitch lag range of 20 to 147 samples.

FSVQ Classifier: FSVQ can be considered as a backward adaptive vector quantizer. It can be viewed as a finite collection of ordinary vector quantizers. For each incoming vector a different codebook is used based on past input vectors. Since the encoder and the decoder both have the same codebook selection procedure, called the next state function, the decoder is able to track the encoder from the past channel symbols --in the absence of channel errors-- without any need for side information.

A speech coder using finite-state vector excitations in the CELP loop can be described as follows: Suppose that we have a state space S and for each member state S_i in S we have a separate quantizer: an encoder γ_{S_i}, a decoder β_{S_i}, and a codebook C_{S_i}. Given a sequence of input vectors $\{X_n; n = 0, 1, 2, \ldots\}$ and an initial state S_0, the channel index vector, the reproduction vector, and the subsequent state label are defined recursively for n=0,1,2,... as

$$V_n = \gamma_{S_n}(X_n), \qquad Y_n = \beta_{S_n}(V_n), \qquad S_{n+1} = f(V_n, S_n) \tag{2}$$

The finite-state classifier can be either a Moore machine (labeled-state) when its outputs are associated with the states, or a Mealy machine (labeled-transition) when its outputs are associated with the transitions [4]. However, we will include only a labeled-state FSVQ in the CELP structure. We have designed our FSVQ classifier in three steps: *1) design the state label codebook, 2) design individual state codebooks, and 3) design the next state function.*

The design of the state-label codebooks is achieved by employing a memoryless VQ design algorithm to obtain one 40-dimensional codeword for each state. Next, using the VQ just designed we have subdivided the residual signal database into a number of partitions, each partition is labeled with a state label identical to the index of the VQ. The second stage is the design of individual state codebooks. Again using the memoryless VQ design algorithm we have designed one codebook for each state using the appropriate portions of the residual signal database. The ensemble of all these codebooks is called the *supercodebook* and we have restricted its size to 512 codewords.

In the third stage, we have chosen the conditional histogram method for its simplicity in computing the next-state function [3]. Assume that the state codebooks are known, then using these codebooks in the CELP synthesis loop of Figure 1 for every training vector of 40 samples, we compute the relative frequency of each codeword given its predecessor. After determining the conditional histogram, the next state function is easily decided by picking the largest occurrence of each codeword with the same predecessor set. This approach was motivated by the observation that in memoryless VQ speech systems, each state was followed almost always by one of a very small subset of states. Thus, the performance should not be impaired if these subsets are fixed as the individual state codebooks.

If the distribution of states is fairly uniform then all the states are equally important. However, if the distribution is highly skewed then it would be difficult to build "good" state codebooks for less probable states. In order to avoid this situation, we have decided to merge less probable states such that the total number of states is fixed and the distribution is fairly smooth. This merging process is completely *ad-hoc* and the specifics of the merging procedure used here are experimentally determined. In our experiments we have used 4, 8, and 16 states and the supercodebook was always fixed to 512 codewords of dimension 40, corresponding to the excitation frame size of 5.0 ms. We have merged a 16-state machine into a system having only four states.

Excitations using Greedy Tree Growing Algorithm: Riskin and Gray have proposed a greedy tree growing algorithm[5] to design codebooks for improving the performance of vector quantizers. The resulting variable rate coders are able to devote more bits to clusters of data that are difficult to code, and fewer bits to less probable data sets. We observed that both the state histograms of the finite-state machine and the relative frequency of codewords in each state codebook display nonuniform character and hence, tree growing algorithms could be exploited to achieve better quality with CELP coders.

To test the above statement we have experimented with a slightly modified version of the original algorithm of Riskin and Gray. They have used $Max\{\frac{\Delta D}{\Delta R}\}$ rule in their splitting mechanism, where ΔD and ΔR correspond to changes in distortion and rate in each leaf, respectively. On the other hand, we have split the cluster with the highest distortion or the most populous one. In order to conform with other systems under consideration each state codebook was limited to 128 codewords.

Synthesis Loop: The FSVQ-CELP synthesizer, lower part of Figure 1, is used just like other CELP coders. However, when the finite-state machine is in the resetting mode -- the system is derailed-- then the index of the minimum weighted distance codeword in the supercodebook is transmitted. Assuming a noiseless channel, the decoder automatically tracks the coder without any error since it has an exact copy of the encoding procedure.

A noise weighting factor of $\gamma = 0.75$ was used for shaping the error signal. In addition, a post-processing procedure consisting of a deemphasis filter of first order with $\mu = 0.75$ and a single-pole single-zero post-filter were employed for enhancing the quality of the synthesized speech.

IMPLEMENTATION AND PERFORMANCE

The speech database was 90 seconds of speech consisting of a number of sentences recorded by a variety of speakers (11 male and 11 female). The speech signals were bandlimited to 100-3600 Hz and sampled at 8 kHz[1]. 640,000 samples from this database were used in designing various codebooks and the remaining 80,000 samples from two male and two female speakers were reserved for testing the system. In initial experiments everything but the LPC coefficients were quantized as described below. Although, there are efficient coding techniques for LPC coefficients at 24 bits per frame[2], this rate was somewhat higher than we could afford for an overall bit rate of 4,000 b/s in our final experiments. Instead a VQ codebook with 512 codewords (9 bits = 450 bits/s) was designed using the Itakura-Saito spectral distortion measure for the short-term linear prediction coefficient sets $\{a_k; (0 \le k \le 10)\}$ obtained from the autocorrelation analysis.

The long-term prediction with quarter sample resolution required 400 bits/s overhead, the pitch lag was coded uniformly into 7 bits $(20 \le k \le 147)$ and pitch predictor coefficients $\{\beta_1, \beta_2\}$ from four consecutive pitch frames were quantized by a 5-bit 8-dimensional VQ. These three components of the long-term predictor needed 400 bits/s, 1,400 bits/s and 250 bits/s, respectively, resulting at a coding rate of 2,050 bits/s.

One particular problem that we have to pay attention closely is when an input vector with low probability goes into the state machine. Since it does not have any "good" reproduction codeword, the FSVQ cannot track the input closely and the system derails. In addition, the accumulation of channel errors can also derail the system. In either case, this problem must be handled by a periodic resetting or by error control or by a combination of the two. We have studied various ways of solving this problem, most of which required additional information to be sent to the decoder. Instead, we have used a simple fixed-time derailing mechanism which did not need any additional bits. After a reasonable number of excitation frames --15 in our final experiments-- we have changed the bit assignment such that the supercodebook consisting of all the codebooks is fully searched. This requires additional 2 bits for a 4-state FSVQ and 4 bits for a 16-state FSVQ, respectively. We have compensated these additional bits by repeating the LPC coefficients of the previous analysis window. There were no objective or subjective noticeable degradations due to this simple technique.

In order to have a total bit rate in the neighborhood of 4,000 bits/s, we decided to use 128-codeword state codebooks for every 5 ms. long excitation vectors. This corresponds to a 1,400 bits/s rate for the excitation vectors. Finally, we have designed a 4-dimensional 32-level codebook for the gain terms adding another 250 bits/s to the total bit rate. Thus, the overall bit rate for all of the coders considered in this study has been limited to 4,150 bits/s.

We have obtained the SNR and segmented SNR values for the proposed systems and those for a reference CELP system operating at the same bit rate. The reference system had one 128-level excitation Gaussian codebook with 4-sigma loading and the rest of the system was identical to the FSVQ-CELP system of Figure 1. These results are tabulated in Table 1.

---

1. This is a copy of the database used by Kroon and Atal [1, page 324]. We would like to acknowledge Peter Kroon for his assistance.

Table 1. Performance of Systems		
System	Uncoded LPC SNR/SegSNR (dB)	Total Rate 4,150 b/s SNR/SegSNR (dB)
CELP	8.37/8.77	7.73/8.36
FSVQ-CELP	8.65/8.35	7.75/8.37
FSVQ-CELP^{GT}	8.03/8.33	7.26/8.13

The SNR and segmental SNR values are very similar for the reference CELP system, the full search FSVQ-CELP, and the unbalanced tree FSVQ-CELP system using greedy tree growing algorithm (FSVQ-CELP^{GT}). The subjective quality of the synthesized speech, however, was slightly better for the proposed FSVQ-CELP^{GT} system. It is worth noting that the quality of the synthesized speech was very similar to that of significantly more complicated systems operating at 4,800 bits/s or higher. In conclusion, the FSVQ-CELP system proposed here can be a viable candidate for half-rate speech coding at 4,000 b/s.

Possible improvement areas are: (1) improved coders for the LPC coefficients [2], (2) savings on pitch-lag bit assignment; (3) structuring the finite-state machine around speech-specific features rather than the histogram-based logic used here, (4) detecting subjectively critical statistical outliers and coding them separately, and (5) replacing the residual codebooks with matching Gaussian codebooks, i.e., design each state codebook from a Gaussian source with statistical and spectral features matching those of the residual signals corresponding to that state.

REFERENCES

[1] B.S. Atal, V. Cuperman and A.Gersho, Editors, *Advances in Speech Coding*, Kluwer Academic Publishers, Boston, MA., 1991

[2] K.K. Paliwal and B.S. Atal, "Efficient Vector Quantization of LPC Parameters at 24 Bits/Frame," in Proc. IEEE Int. Conf. Acoust., Speech, Signal Processing, pp. 661-664, May 1991, Toronto, Canada.

[3] M. O. Dunham and R.M. Gray, "An Algorithm for the Design of Labeled-Transition Finite-State Vector Quantizers," IEEE Trans. on Communications, Vol. COM-33, No. 1, January 1985.

[4] R.M. Gray, "Vector Quantization," IEEE Acoustics, Speech and Signal Processing Magazine, April 1984, also in *Vector Quantization*, H. Abut, Ed., IEEE Press, New York, N.Y., 1990.

[5] E.A. Riskin and R.M. Gray, "A Greedy Tree Growing Algorithm for the Design of Variable Rate Vector Quantizers," IEEE Trans. on Signal Processing, vol. ASSP-39, pp. 2500-2507, November 1991.

AUTHOR INDEX

Abut, H. 271
Adoul, J-P. 147
Atal, B. S. 191, 239
Barnwell, III, T.P. 19, 259
Benyassine, A. 271
Bhaskar, U. 265
Champion T.G. 127
Chan, W.-Y. 153
Chen, J-H 25
Cox, N. B. 203
Cuperman, V. 33, 101
De Iacovo, R. D. 141
de Marca, J.R.B. 163
De Martino, E. 55
Delprat, M. 93
Dervaux, F. 93
Dimolitsas, S. 43
Dymarski, P. 231
Farvardin, N. 181
Foodeei, M. 5
Fuldseth, A. 121
Gardner, 85 W.
Gersho, A. 77, 153, 251
Gerson, I. A. 211
Granzow, W. 111
Grass, J. 5
Gruet, C. 93
Gupta S. K. 239
Hansen, H.B. 171
Harborg, E. 121
Husain, A. 33
Jacobs, P. 85
Jasiuk, M.A. 211
Johansen, F.T. 121
Kabal, P. 5
Kataoka, A. 11
Kleijn, B.A. 111
Knudsen, J.E. 121
Laflamme, C. 147
Larsen, K.J. 171
LeBlanc, W. 101
Lee, C. 85
Mahmoud S.A. 101
Marques, G.C. 271
Mazor, B. 225
McAulay, R.J. 127
McCree A.V. 259
Mermelstein, P. 5, 69
Mikkelsen, K.B. 171
Montagna, R. 141
Moreau, N. 231
Moriya, T. 11, 181
Nayebi, K. 19
Nielsen, H.J. 171
Ohta, Y. 217
Paksoy, E. 77, 251
Paliwal, K.K 191
Panzer, I.L. 59
Phamdo, N. 181
Quatieri T.F. 127
Rauchwerk, M.S. 25
Salami, R. 147
Sereno, D. 141
Sharpley, A.D. 59
Shoham, Y. 133
Su, H-Y 69
Tanaka, Y. 217
Taniguchi, T. 217
Usai, P. 141
Veeneman, D. 225
Voiers, W.D. 59
Wang, S. 251

INDEX

APC-TQ, 265
Absolute Category Rating (ACR), 48, 59
Adaptive codebook, 37, 212
Adaptive rate decision, 89
Adaptive spectral enhancement, 262
Algebraic CELP (ACELP), 147
Algebraic codebooks, 150
Algorithmic delay, 17, 33
Analysis-by-synthesis, 121, 173
Analysis-synthesis systems, 19
Aperiodic pulses, 261
Audio coding, high fidelity, 153
Average rate, 85
Backward LPC analysis, 134
Backward adaptive gain, 14
Backward-adaptive CELP, 11
Backward-adaptive LPC predictor, 26
Bit interleaving, 172
Bit-error sensitivity, 164
Block release, 136
Blockwise interpolation, 112, 247
Branching factor, 6
Burst-error-correcting codes, 172
CCITT G.728 standard, 25
CELP, 5, 79, 86, 121, 141, 196, 211, 231, 271
Channel coders, 174
Channel errors, 194
Channel optimized VQ, 168
Channel-matched MSVQ (CM-MSVQ), 182
Closed-loop analysis, 226
Closed-loop training, 28
Code Division Multiple Access (CDMA), 83, 85
Codebook adaptation, 220
Codebook design, 38
Coding delay, 73
Conditional pitch prediction, 11
Constrained excitation, 97
Constrained storage VQ (CSVQ), 155
Critical bands, 19, 239
DAM test scores, 263
DAM, 59

DCR, 61
DCT, 266
DMOS, 59
Degradation Category Rating (DCR), 48, 61
Delayed-decision coding (DDC), 5, 71, 133
Delta codebook, 217
Delta pitch encoder, 37
Delta vector sorting, 221
Delta vector, 218
Derailing algorithm, 271
Derailing effect, 138
Diagnostic Acceptability Measure (DAM), 59
Digital cellular standard(s), 55
Digital conferencing, 127
Directed tree, 135
Dynamic multirate, 129
Enhanced TDMA (E-TDMA), 82
Equality Threshold Rating (ETR), 48
Error control mapping, 204
Estimated residual method, 227
European digital mobile radio, 93
Excitation gain control, 97
Excitation model, 271
FIR approximation, 36
FSVQ classifier, 272
Fading, 171
Finite-state VQ (FSVQ), 271
Fixed delay level coding, 73
Forward error control (FEC), 204
Frame lag trajectory, 211
Frequency domain representation, 240
Frequency-dependent voicing, 259
G.711, 268
G.712, 269
G.721, 265
G.722, 141
GSM half rate channel, 93
GSM, 55
Gain adaptation, 28
Gain predictor, 28
Generalized Analysis-by-Synthesis, 117
Generalized Lloyd algorithm, 12, 103
Generalized product codes (GPC), 153
Generalized pseudo-Gray coding, 203
Greedy tree growing, 271, 272

Group Special Mobile (GSM), 93
Half-rate GSM, 171
Harmonic noise weighting, 213
IS-54, 163
Impairments, 44
Incremental release, 136
Index assignment optimization, 203
Index assignment, 164
Interframe predictive coding, 27
Interpolated lag prediction filters, 227
Joint minimization, 71
LBG algorithm, 167
LPC excitation, 239
LPC quantization, 191, 251
LPC residual coding, 69
LPC vocoders, 259
LSF quantization, 252
LSP parameter weighting, 187
LSP quantization, 141
LTP, 211
Lattice low-delay VXC (LLD-VXC), 34
Line spectral frequency (LSF), 251
Line spectrum pair (LSP), 181, 191
Listener opinion test, 43, 47
Long-term predictor, 211, 225
Low Delay Coding of Wideband Speech, 133
Low delay speech coder, 11
Low delay, 33
Low-delay CELP(LD-CELP), 25, 133
Low-delay coding, 25
Low-delay filter banks, 19
Low-delay subband coder, 22
M-L tree search procedure, 102
ML-Tree algorithm, 5
Masking threshold, 154
Mean Opinion Score (MOS), 46, 56, 59, 80, 81
Mixed excitation, 259
Modulated Noise Reference Unit (MNRU), 46, 56, 62, 95, 144
Multi-speaker conferencing, 129
Multi-stage VQ, 101
Multi-stage vector quantization (MSVQ), 182
Multi-tap pitch prediction, 225
Multirate STC, 128
Multistage CELP, 231
Multistage vector quantization (MSVQ), 251

Next state function, 273
Noise weighting filter, 121
Non-integer pitch lags, 227
Non-linear interpolative VQ (NLIVQ), 156
Non-uniform filter banks, 20
Non-uniform frequency domain sampling, 239
Orthogonalization, 233
Packet reservation multiple access, 83
Parametric representation of the LPC excitation, 239
Partially-forward system, 38
Path map, 6
Path merging effect, 138
Perceptual distortion, 154
Pitch cycle waveform, 111
Pitch prediction, 225
Pitch search complexity, 150
Pitch-adaptive excitation, 14
Predictor adaptation, 36
Product-code VQ, 252
Prototype waveform interpolation (PWI), 112
Prototype waveforms, 111
Pseudo-Gray coding, 27
Punctured convolutional codes, 173
QCELP, 79, 85
QR Factorization, 231
RPE-LTP, 55
Reed-Solomon (RS) codes, 172
Real-time implementation, 151, 262
Regular Pulse CELP, 93
Regular Pulse Excitation Long Term Prediction (RPE-LTP), 55
Regular pulse codebook, 96
Robustness to channel errors, 203
SVQ, 251
Search complexity, 106
Signal to Change Ratio (SCR), 113
Simultaneous joint codebook design, 103
Single pulse codebook, 96
Sinusoidal Transform Coder (STC), 128
Source-channel coding, 181
Spectral distortion, 102, 251
Speech coder quality, 43
Speech quality evaluation, 55
Speech quality, 59
Split VQ, 252
Split vector quantization, 251

Split-band CELP, 145
Standardization, 43
Structured VQ, 153
Subjective assessment methods, 43
Subjective testing, 55, 56
Supercodebook, 273
Switched-adaptive vector predictor, 253
TIMIT database, 108
Three-tap pitch predictor, 27, 225
Time-windowed basis functions, 240
Tree structured VQ (TSVQ), 156
Tree-coding, 69
Tree-delta codebook, 218
Tree-searched MSVQ, 107
Trellis diagram, 155
Unequal error protection, 163
VQ, 166
Variable rate CELP, 80
Variable rate coder, 85
Variable rate speech coders, 77
Variable-rate transform coding, 158
Vector Sum Excited Linear Prediction (VSELP), 55, 215
Vector linear prediction, 252
Vector quantization (VQ), 122, 191, 203, 267
Vector tree CELP, 133
Videophone, 121
Viterbi decoding, 173
Voice activity factor, 78
Voice activity, 78
Waveform coder, 111
Waveform matching, 111
Weighted LSF distance measure, 192
Wideband speech coder, 133, 147
Wideband speech, 121
Zero-redundancy channel coding, 164